Amazing Latest Science Discoveries

Fun Facts Book For Adults

Brilliant, mindblowing and downright bizarre new discoveries in Science, Health, History and Nature that are changing our world.

Third Rock Publishing

First edition 2025

Third Rock Publishing
Email: info@thirdrockpublishing.com

EDITOR'S NOTE

Welcome, fellow explorer!

At Third Rock Publishing, our team is driven by a passion to bring you the most astonishing, up-to-date, and conversation-starting discoveries from the world of science. This book is the result of months of deep research, fact-checking, and genuine curiosity about our ever-changing universe.

We hope these pages inspire wonder, spark new questions, and maybe even change the way you see the world. Every fact has been carefully chosen and linked to an original source, usually one of many, because we believe you deserve the truth, not just trivia.

Thank you for joining us on this adventure. Stay curious!

Maggie

Editor

Third Rock Publishing
Email: info@thirdrockpublishing.com

TABLE OF CONTENTS

Weird Science Facts: Unveiling the Latest Wonders and Whimsies

These latest science discoveries aren't just quirky—they're changing how we understand our bodies, our planet, and the universe itself. Dive in and explore the delightful, sometimes bizarre, and always fascinating side of science that proves reality is often wilder than our wildest imagination.

Here you will meet cows learning bathroom manners, glowing sharks, trees whispering through underground networks, and brains with hidden control panels.

1

Your Brain Has a Built-In Emotion Control Panel

Ever wish you had better control over your emotions? Turns out, your brain already has the tools—it just took scientists a while to find them. In 2024, researchers at Dartmouth mapped out the brain's emotion regulation systems, separating the circuits that feel emotions from those that manage them. It's like discovering your brain has both a thermostat and an air conditioner. These insights could lead to smarter therapies for mental health and addiction—no rewiring required.

Source:
https://www.sciencedaily.com/releases/2024/04/240403170916.htm

2

The Amazon's Carbon Reversal Problem

Historically hailed as the "lungs of the Earth" for its role in absorbing vast amounts of carbon dioxide, the Amazon Rainforest is now a carbon emitter instead of a carbon sink. Climate change and deforestation have transformed it into a net source of carbon. It's a troubling turn for an ecosystem that supports more than a third of the world's species. Conservationists hope this "alarm bell" could prompt global action to preserve and regenerate these ancient trees before their storage capacity is permanently compromised.

Source:
https://www.theguardian.com/commentisfree/2021/jul/19/amazon-deforestation-jair-bolsonaro-rainforest-carbon-contributors

3

Potty-Trained Cows Could Help the Climate

Researchers discovered in 2024 that cows could be trained to use designated bathroom areas. In a unique twist on house-training a pet, scientists successfully taught cows to "go" in certain areas, reducing ammonia emissions from manure that contribute to climate change. It turns out these animals are smart enough to catch on to positive reinforcement techniques, which could help farming become more eco-friendly. Who knew cows could be toilet-trained for a greener planet?

Source:
https://www.thebrighterside.news/post/scientists-are-toilet-training-cows-to-save-the-planet-2/

4

Space Anemia: A Surprising Challenge for Astronauts

A 2024 study revealed that astronauts experience accelerated loss of red blood cells in space, leading to what scientists are calling "space anemia." This condition arises from the unusual effects of low gravity on the body and represents a big challenge for long-term space travel. In space, blood cells are destroyed twice as fast as on Earth, which could mean scientists will need to develop new health protocols for astronauts, especially for extended missions to Mars or beyond.

Source:
https://www.sciencedaily.com/releases/2022/01/220114115708.htm

5

Trees' Underground "Internet" is Like the Original Wi-Fi

Did you know trees "talk" to each other? Through a complex underground network of fungi, plants share nutrients, warn of insect attacks, and even "gossip" about environmental changes. Known as the "Wood Wide Web," this network supports ecosystem health and resilience, a fact confirmed by 2024 research. Imagine trees sending each other warnings and instructions as a way to look out for one another, with scientists now working on applying these findings to agricultural techniques.

Source:
https://www.newscientist.com/article/mg25333720-500-how-the-world-really-works-review-the-tech-that-underpins-society/

6

Glowing Sharks are the Ocean's Brightest Secret

Researchers in 2024 uncovered that several deep-sea shark species emit a natural glow. This light-up effect, called bioluminescence, might help these sharks attract mates or ward off predators in the darkness of the deep ocean. They control this glow using a natural "switch," showcasing an impressive adaptation that deepens our understanding of how animals evolve to survive and thrive in extreme environments.

Source:
https://www.smithsonianmag.com/smart-news/nearly-six-foot-glowing-shark-discovered-deep-sea-new-zealand-180977163/

7

Your Emotions Might Soon Be Graded by AI - in a Good Way

Emotion AI has emerged as a transformative technology with significant potential to improve healthcare and education.

Using data from your facial expressions, voice tone, body language, and even brain signals, this next-gen tech can detect your emotions in real time. Imagine a classroom where teachers get instant emotional feedback on who's engaged and who's zoning out, or doctors who can track a patient's mental well-being just by reading subtle changes in expression.

The goal? More personalized care and smarter learning.

Source:
https://www.frontiersin.org/journals/psychology/articles/10.3389/fpsyg.2024.1533053/full

8

Octopuses Might Dream in Technicolor

A recent study suggested that octopuses might experience vivid dreams that change their skin color while they sleep. Researchers observed color-shifting patterns on sleeping octopuses, hinting that they may be dreaming, similar to how humans experience rapid eye movement (REM) sleep. This discovery makes us wonder: if an octopus can dream, could they be imagining swimming free or hunting for tasty treats in the ocean depths?

Source:
https://www.newscientist.com/article/2272319-octopuses-may-be-able-to-dream-and-change-colour-when-sleeping/

9

AI Detects Lies Better Than People Can

An AI-based lie detector developed in 2024 has achieved an impressive 92% accuracy rate in detecting lies by analyzing body language and facial cues. While human accuracy for spotting deception typically hovers around 54%, AI could change how we think about trust in high-stakes situations like courts or interrogations. Will this mean a future where our "little white lies" don't stand a chance?

Source:
https://www.bloomberg.com/news/articles/2024-09-05/ai-startup-wants-to-help-you-detect-when-someone-is-lying-at-work

10

Printing Human Organs in Space Could Be a Reality

In 2024, scientists achieved new breakthroughs in printing human tissues in space. Microgravity conditions seem to make the 3D printing of human organs and tissues easier, allowing researchers to explore ways of creating medical supplies in orbit. This discovery could be revolutionary for future space travelers, who might one day have emergency organ replacements printed out while on a mission to Mars or beyond.

Source:
https://www.drugdiscoverynews.com/3d-bioprinting-tissues-in-space-to-heal-people-on-earth-1-15845

11

THE HEART HAS AN ENERGY FIELD

In 2024, scientists found more evidence that the heart produces a powerful electromagnetic field. This field extends several feet beyond the body, influencing how our bodies interact with the world and other people around us.

Your heart isn't just beating—it's broadcasting energy. Ancient cultures always believed the heart was more than a pump, and science is now catching up.

Source:
https://www.rfsafe.com/heart-to-brain-communication-rhythms-within-the-human-electromagnetic-field/

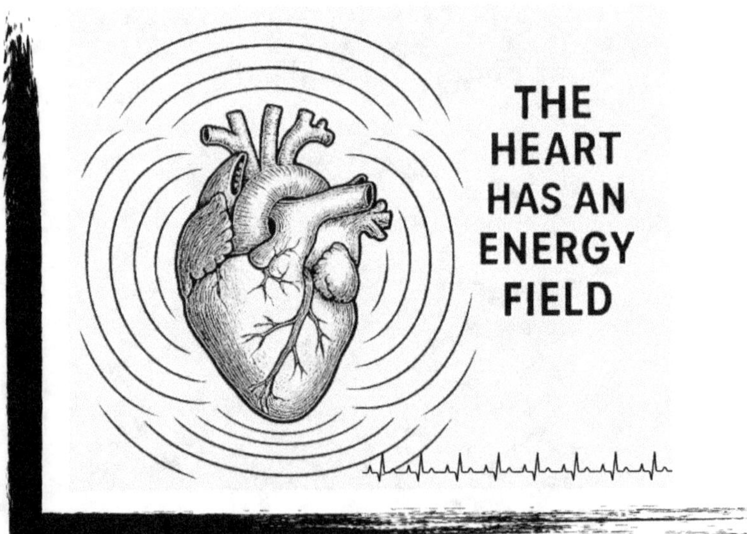

12

BRAIN CELLS CAN REGROW AFTER INJURY

Doctors used to think brain cells couldn't regrow, but in 2024, researchers confirmed that certain areas of the brain can regenerate after injury. This discovery could lead to breakthroughs in treating brain damage and strokes.

Your brain's healing powers are far greater than anyone realized. It's like having a hidden superhero up there, ready to spring into action.

Source:
https://www.sciencedaily.com/releases/2024/10/241017112725.htm

13

THE EARTH'S CORE IS SLOWING DOWN

In early 2024, geophysicists discovered that the Earth's inner core is rotating more slowly than its outer layers. While this isn't dangerous, it does affect the length of our days, changing them ever-so-slightly.

Our planet's core is basically a gigantic, slowly ticking clock, and it's still throwing surprises our way.

Source:
https://phys.org/news/2024-06-rotation-earth-core.html

14

Bacteria Are Thriving in Your Showerhead

New research in 2024 shows that biofilms—groups of bacteria—can form in showerheads and then spray bacteria-laden water on you. While most of these bacteria are harmless, some can cause illness in people with weakened immune systems.

It's like a bacteria party in your shower. Not exactly the party you want to join, but science is here to help you keep it in check.

Source:
https://www.frontiersin.org/journals/microbiomes/articles/10.3389/frmbi.2024.1416055/full

15

Tree Rings Show Ancient Climate Change Patterns

A 2024 study of tree rings from thousands of years ago reveals a detailed picture of climate patterns that ancient civilizations lived through. This research could help predict future climate shifts. Trees are like nature's historians, quietly recording our planet's climate secrets in their rings.

Source:
https://phys.org/news/2024-07-tree-reveal-north-atlantic-sea.html

Brain Bites - New Fun Facts in Neuroscience

Here are 10 fun and quirky neuroscience facts, with new research and unexpected twists to intrigue you in a fresh, memorable way.

1

Your Brain's Storage Power Rivals Giant Data Centers

The human brain is estimated to store up to 2.5 petabytes of information—roughly equivalent to 3 million hours of TV shows. While it's not directly comparable to the internet's size (measured in zettabytes), this makes the brain one of the most efficient storage systems known. Instead of saving everything, it optimizes by strengthening important connections and pruning less-used ones, allowing it to adapt, learn, and recall with remarkable speed and flexibility.

Source:
https://www.intelligentliving.co/human-brains-data-storage-10-times/

2

Your Brain Is a Sugar-Hungry Machine

Despite making up only about 2% of your body weight, your brain consumes around 20–25% of your body's energy - most of it in the form of glucose, its primary fuel. Glucose comes from the breakdown of carbohydrates in your diet. Foods high in simple sugars like honey, fruit, or table sugar provide a quick source of glucose, leading to a rapid rise in blood sugar and fast energy for your brain. In contrast, complex carbohydrates found in whole grains, legumes, and starchy vegetables are broken down more slowly, giving your brain a steadier, longer-lasting supply of glucose.

So, whether you eat honey on toast, a bowl of oatmeal, or a piece of fruit, you're fueling your brain's remarkable energy needs.

Source:
https://pmc.ncbi.nlm.nih.gov/articles/PMC11876560/

3

Yawning - Your Brain's Built-In Cooling System

Yawning isn't just a sign of boredom or fatigue — it's a biological way to cool your brain. Studies suggest that yawning helps regulate brain temperature, especially in mammals, whose longer yawns are linked to larger brain sizes and slower blood cooling compared to birds. Remarkably, just watching someone yawn—even in a video—can trigger your own, thanks to automatic activity in the brain's motor regions. Some scientists believe yawning may also serve as a subtle social cue to signal alertness or a need for group vigilance, though the role of empathy is still being explored. And if you've ever struggled to suppress a yawn, that's because your brain is wired to make it nearly impossible.

Source:
https://phys.org/news/2022-06-evolutionary-biologist.html
https://www.sunypoly.edu/news/news-release-longer-yawn-bigger-brain-utrecht-university-suny-poly-research-published.html

https://sleepreviewmag.com/sleep-disorders/hypersomnias/yawning-cools-brain/

4

Your Brain Makes Backup Copies of Your Memories

Think your brain is like a messy drawer of random thoughts? Think again. New research shows that your brain actually stores at least three copies of every memory like a computer. These copies allow you to recall details, adapt memories over time, and learn new things without getting overwhelmed.

Scientists also found a special "molecular glue" called KIBRA that helps lock memories in place, and that sleep resets your brain each night, preparing it to store more tomorrow. So yes — that power nap really is helping you remember where you left your keys.

Source:

https://www.sciencedaily.com/releases/2024/08/240815163622.htm

https://neurosciencenews.com/genetics-memory-recall-26377/

5

Music Can Supercharge Your Brain's Healing Powers

Listening to music can rebuild brain connections after injury, like a stroke. In one study, patients who listened to their favorite music recovered speech and memory abilities faster than those who didn't. When combined with magnetic stimulation (TMS) the results are even better. Music doesn't just make you feel good—it literally helps your brain heal.

Source:
American Academy of Neurology

https://www.neurology.org/doi/10.1212/WNL.0000000000212171

6

The Brain Treats Heartbreak Like Physical Pain

Heartbreak doesn't just "feel" painful—it lights up the same parts of the brain that process physical pain. So when people say their heart is "aching" after a breakup, their brain is genuinely reacting as if they'd been physically hurt.

Source:
https://dreliaz.org/study-shows-heartbreak-can-hurt-physically/

7

You Can "Erase" and "Overwrite" Old Memories

Scientists have found that certain memories can be rewired or "overwritten" with new ones through a process called reconsolidation. It's like editing an old file on your computer—your brain can update your memory with new information, which can even help treat PTSD. By reactivating traumatic memories in controlled settings, therapists can help patients modify the emotional intensity associated with those memories, leading to reduced symptoms.

Source:
https://www.nature.com/articles/s41386-024-02049-0

8

Your Brain Has a Night Shift Cleaner

Scientists have discovered that your brain has its own secret cleaning system called the glymphatic system—and it only works well while you're asleep. This system flushes out waste and toxins from your brain by circulating cerebrospinal fluid through special channels. It's like a night janitor for your brain, helping to prevent buildup that's linked to conditions like Alzheimer's.

Thanks to advanced MRI scans and special tracers in neurosurgery, researchers have now seen this system in action in living humans for the first time. They found that good, deep sleep supercharges the process, while poor sleep slows it down — meaning your brain's mess piles up. The discovery not only confirms the system's existence in people but also highlights just how crucial quality sleep is for protecting your brain's long-term health.

Source:

https://news.ohsu.edu/2024/10/07/brains-waste-clearance-pathways-revealed-for-the-first-time

https://www.nih.gov/news-events/nih-research-matters/brain-waste-clearance-system-shown-people-first-time

https://interhospi.com/brains-waste-clearance-system-definitively-imaged-in-humans-for-the-first-time/

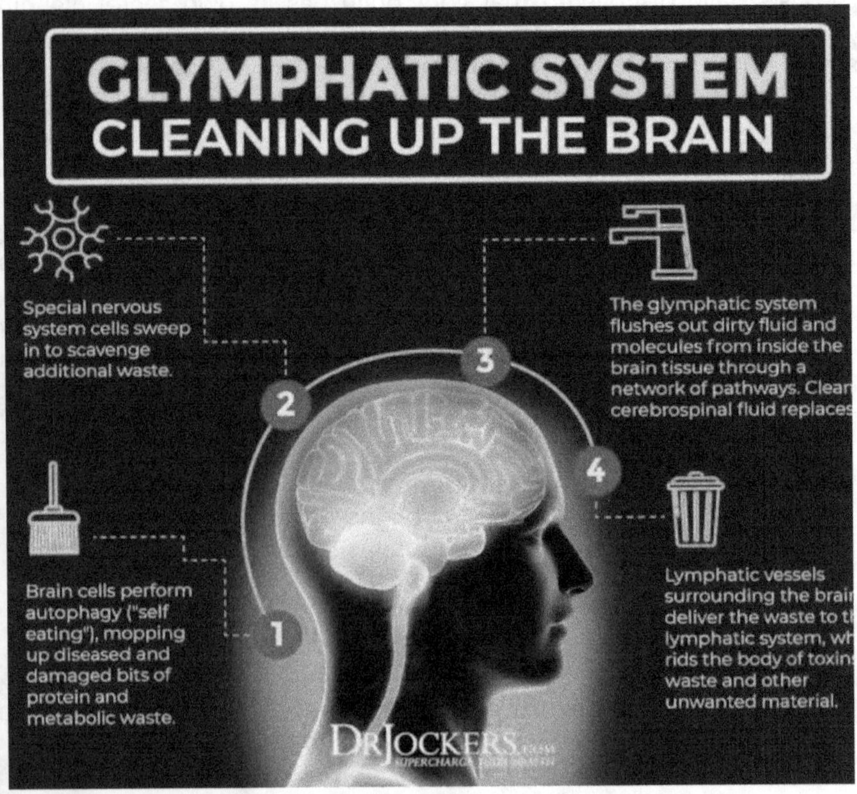

Image credit:
https://drjockers.com/

9

Meditation Can Physically Reshape Your Brain

Long-term meditation has been shown to increase the size of the hippocampus (linked to memory and learning) and shrink the amygdala (associated with fear and stress). Essentially, you can sculpt your brain through mindfulness—like a mental workout.

Source:
https://giving.massgeneral.org/stories/meditation-reduces-stress-anxiety

10

The "Gut Feeling" Is Real—Your Brain Talks to Your Gut

The "gut feeling" you sometimes get isn't just in your head - it's in your gut too. Recent Johns Hopkins Medicine research confirms that your "gut feeling" is real: the brain and digestive system communicate constantly via the vagus nerve and the enteric nervous system (sometimes called the second brain). This explains why emotions and stress can affect digestion and why gut instincts are rooted in real biology.

Source:
https://www.hopkinsmedicine.org/health/wellness-and-prevention/the-brain-gut-connection

Quiz: New in Neuroscience

Think you know your brain? Put your mind to the test with these 10 fascinating neuroscience questions!

1. **How much storage capacity does the human brain have?**
   a. About 1 terabyte
   b. About 2.5 petabytes
   c. About 100 gigabytes
   d. Infinite

2. **What surprising way can yawning be contagious?**
   a. Through direct eye contact only
   b. Only when hearing someone yawn
   c. By seeing a photo or video of someone yawning
   d. It's not contagious—it's just a myth

3. **How much of your body's energy does your brain consume?**
   a. 2%
   b. 10%
   c. 20%
   d. 50%

4. **What happens to your memories each time you recall them?**
   a. They stay exactly the same
   b. They slowly fade away
   c. They get slightly reconstructed and can change
   d. Your brain deletes the old ones to make space for new ones

5. **How can music help the brain heal after an injury?**
   a. It improves hand-eye coordination
   b. It helps reconnect damaged brain pathways
   c. It strengthens the immune system
   d. It makes people forget their injury

6 **Why does heartbreak feel physically painful?**

a. It weakens the heart muscle
b. It affects the same brain regions as physical pain
c. It decreases oxygen flow to the brain
d. It's just an emotional reaction, not a physical one

7 **What is "reconsolidation" in neuroscience?**

a. A process that strengthens old memories over time
b. A way for the brain to erase and update old memories
c. The brain's ability to create new neurons
d. The process of storing memories in the spinal cord

8 **What does the brain do while you sleep, thanks to the glymphatic system?**

a. Stores new information permanently
b. Repairs damaged neurons
c. Clears out waste and toxins
d. Increases blood circulation

9 **How does meditation change the brain?**

a. It makes people smarter
b. It shrinks the amygdala (fear center) and grows the hippocampus (memory center)
c. It slows down brain activity permanently
d. It makes the brain immune to stress

10 **What connects your brain to your gut, allowing them to communicate?**

a. The vagus nerve
b. The spinal cord
c. The digestive tract
d. The brainstem

ANSWERS

1. b) About 2.5 petabytes
2. c) By seeing a photo or video of someone yawning
3. c) 20%
4. c) They get slightly reconstructed and can change
5. b) It helps reconnect damaged brain pathways
6. b) It affects the same brain regions as physical pain
7. b) A way for the brain to erase and update old memories
8. c) Clears out waste and toxins
9. b) It shrinks the amygdala (fear center) and grows the hippocampus (memory center)
10. a) The vagus nerve

The Science Lab Inside Your Head: Quirky Facts About How Our Brains Really Work

Our brains are full of surprises — from playing tricks with our memories to sparking bursts of creativity or even nibbling away at themselves.

These fascinating facts that reveal the strange, brilliant, and downright amazing ways our minds operate, straight from the world of neuroscience and psychology.

1

You Can "Hack" Your Brain Into Forgetting Some Things

What if you could train your brain to forget unwanted memories? Sounds like science fiction-but it's real, thanks to a phenomenon called retrieval-induced forgetting (RIF). Researchers have discovered that when you deliberately focus on recalling some memories, your brain actively suppresses related-but-unpracticed memories, making them fade away. It's like your brain's own selective "delete" button!

Cutting-edge studies show this isn't just a fleeting trick. In carefully controlled experiments, people who practiced retrieving certain words or facts ended up forgetting others they'd learned but didn't practice. This effect is so robust that scientists have observed it across different ages, types of memories, and even emotional content. The brain's ability to inhibit competing memories is a powerful tool for managing what we remember - and what we let go.

A landmark 2025 study revealed that while young adults can "release" these suppressed memories over time, older adults experience longer-lasting forgetting, highlighting how memory control changes with age. Cognitive scientists continue to unravel how this fascinating process shapes our memory every day.

Source:
https://pmc.ncbi.nlm.nih.gov/articles/PMC11939593/

2

The Brain Can "Delete" Information During Sleep

Ever wonder why a good night's sleep helps you feel refreshed? During sleep, your brain is busy clearing out unimportant information to make room for new learning. This "mental housekeeping" helps sharpen your focus the next day, but it also means some details might disappear. It's like your brain is taking out the trash while you dream.

Source:
https://www.nature.com/articles/s41467-025-58860-w

3

Your Brain "Eats" Itself When You're Tired

Sleep deprivation can cause your brain cells to get so confused that they start devouring parts of themselves—literally. This self-cannibalism happens to clean out old, damaged cells, but when you're tired, the process can go into overdrive. So next time you feel foggy after an all-nighter, it might be your brain eating itself to stay awake.

Source:
https://medicalxpress.com/news/2025-01-powering-brain-health-autophagy-neurons.html#google_vignette

4

YOUR FAVORITE SONGS LIGHT UP YOUR BRAIN

When you hear a favorite song from your past, such as the anthem from your first school dance or a tune from a family road trip, your brain does more than just remember.

Recent brain imaging studies show that nostalgic music lights up a unique network of regions tied to memory, emotion, and self-reflection, including the default mode network and the brain's reward centers. These songs evoke vivid autobiographical memories and can even boost mood and cognitive function, especially in older adults. That's why a single melody can instantly transport you back in time, making nostalgia not just a feeling but a powerful brain event.

Source:
https://medicalxpress.com/news/2025-04-nostalgic-music-minds.html#google_vignette

5

Mind-Reading Is Real

Scientists have cracked the code to literally see what's inside your head. Using super-smart AI and brain scans, researchers can now turn your brainwaves into actual images. Yes, your thoughts, what you're looking at, even your dreams decoded and recreated.

In 2024, a team at Radboud University blew minds by teaching AI to read brain signals and reconstruct faces and objects almost perfectly. It's like your brain's secret photo album is finally getting exposed!

This breakthrough could revolutionize medicine, communication, and how we understand consciousness itself. So, next time you daydream, remember: your brain might be broadcasting more than you think.

Source:
https://petapixel.com/2024/08/05/this-ai-recreates-images-from-brain-waves-more-accurately-than-ever-before/

6

Your Brain Rewrites History (Every Time You Recall a Memory)

Every time you recall a memory your brain slightly alters it. That's right - your brain is like a storyteller that edits the narrative each time it's retold. This is why two people can have totally different versions of the same event. Your memories are more like a movie script in constant revision than a photograph frozen in time.

Source:
https://cdn.fortunejournals.com/articles/reconsolidation-behavioral-updating-of-human-emotional-memory.pdf

7

Daydreaming Makes You More Creative

If you've ever been caught staring off into space, don't feel guilty. Daydreaming is actually good for your brain. Studies show that when you let your mind wander, it taps into your default mode network, which helps boost creativity and problem-solving. So, zoning out can sometimes lead to breakthroughs.

Source:
https://www.nature.com/articles/s42003-025-07470-9

8

The Brain Can "Feel" a Fake Limb

In a peculiar phenomenon known as the rubber hand illusion, your brain can be tricked into feeling a fake limb as if it were your own. Simply by seeing a rubber hand being touched while your real hand is hidden, your brain adopts the fake one as part of your body. This shows how easily the mind can be fooled into believing what it sees.

Source:
https://neurosciencenews.com/rubber-hand-pain-28492/

9

You Can Change Your Brain Structure by Learning

Your brain isn't fixed—it's constantly rewiring itself based on your experiences. This is called neuroplasticity, and it means that learning new things, like a language or a musical instrument, physically changes your brain. The more you practice, the stronger the connections between neurons become. Your brain is like a living sculpture that keeps reshaping itself.

Source:
https://medicalxpress.com/news/2024-05-mechanism-neural-plasticity-underlying-memory.html

10

Laughter Really Is the Best Medicine

When you laugh, your brain releases endorphins—the body's natural painkillers—and it even boosts your immune system. Laughter helps to lower stress hormones and increases the number of immune cells and antibodies in your body. No wonder laughter is contagious—it's a brain-boosting superpower.

Source:
https://www.mayoclinic.org/healthy-lifestyle/stress-management/in-depth/stress-relief/art-20044456

11

Your Brain Has a Cellular "ID Card"

Recent research has revealed that each brain region possesses a unique cellular signature, influenced by sensory experiences. This means that activities like seeing and hearing shape the development of specific brain areas, effectively giving each region a distinct "ID card."

Source:
https://news.mit.edu/2023/brain-self-supervised-computational-models-1030?

12

Advancements in Alzheimer's Treatment

Innovative treatments are being explored to combat Alzheimer's disease.??One such approach involves using brain wave stimulation to reduce amyloid plaques, which are associated with the disease. This method has shown promise in slowing down brain degeneration.

Source:
https://news.mit.edu/2023/new-wave-treatment-alzheimers-disease-li-huei-tsai-1115?

13

Rapid and Precise Brain Mapping

Scientists have developed advanced tools that enable the mapping of millions of neurons simultaneously. Techniques like BARseq allow for the rapid creation of detailed brain maps, enhancing our understanding of neural connections and their evolution over time.

Source:
https://news.mit.edu/2023/using-ai-optimize-rapid-neural-imaging-1106?

14

Brains Generate Their Own "Music"

Neurons in the brain can synchronize in rhythmic patterns, akin to musical beats. These natural rhythms play a crucial role in regulating functions such as movement and memory, continuously operating in the background of our minds.

Source:
https://www.livescience.com/health/neuroscience/18-brain-studies-that-blew-our-minds-in-2023?

15

Gut Bacteria Influence Mood

Studies have found that certain gut bacteria can affect the brain's mood centers. These microbes produce chemicals that travel through the bloodstream to the brain, potentially explaining sensations like "butterflies" before a significant event.

Source:
https://www.npr.org/sections/shots-health-news/2024/06/24/nx-s1-5018044/gut-microbiome-microbes-mental-health-stress

16

Adolescent Brain "Pruning"

During adolescence, the brain undergoes a significant reorganization, eliminating unnecessary neural connections to enhance efficiency. This "pruning" process is akin to a gardener trimming a tree, resulting in sharper cognitive functions as individuals mature.

Source:
https://raisingchildren.net.au/pre-teens/development/understanding-your-pre-teen/brain-development-teens

17

Brain Waves Control Prosthetic Limbs – With Feeling.

Imagine moving a robotic arm just by thinking about it-and actually feeling what it touches. Thanks to groundbreaking brain-computer interface (BCI) technology, this is no longer science fiction. Tiny electrodes implanted in the brain or nerves read your thoughts and translate them into natural, fluid movements of prosthetic limbs. Even more astonishing, new systems send sensory signals back to the brain, letting users "feel" textures and shapes through their artificial hands.

This leap in neuroprosthetics is restoring independence and transforming lives. People with limb loss can now perform delicate tasks-gripping a coffee cup, typing on a keyboard, or climbing stairs - with precision and ease. By 2030, these mind-controlled prosthetics will become even more intuitive, widely available, and capable of mimicking the full experience of natural limbs.

Source:
https://www.robobionics.in/blog/neural-controlled-prosthetics-how-brain-integrated-tech-is-changing-lives-2025-2030/

18

Rewriting the Past to Heal the Future

Imagine if you could erase or soften painful memories. Scientists are getting closer to making that a reality. By understanding exactly how memories are stored and recalled in the brain, researchers are developing ways to target and alter specific neural pathways. This means it might soon be possible to weaken or even erase traumatic memories, offering new hope for people suffering from PTSD and other mental health conditions. The brain's ability to rewrite its own history could transform how we heal emotional wounds.

Source:
https://www.frontiersin.org/journals/neurology/articles/10.3389/fneur.2024.1481450/full

19

Sleep "Cleans" the Brain

During sleep, the brain engages in a cleansing process, removing toxic proteins that accumulate during wakefulness. This nightly detoxification underscores the importance of sleep for maintaining optimal brain health.

Source:
https://medicine.washu.edu/news/neurons-help-flush-waste-out-of-brain-during-sleep/

20

The Brain's Internal "GPS"

A specialized group of neurons, known as "place cells," function as the brain's internal GPS. These cells help us navigate and remember locations, shedding light on why some people have a more innate sense of direction than others.

Source:
https://neurosciencenews.com/sleep-glympthatic-norepinepherine-28332/

Quiz: Brain Benders

Multiple Choice (1–10)

Circle the best answer for each question.

1 **What is the term for training your brain to forget unwanted memories?**

a. Selective memory loss
b. Memory dumping
c. Retrieval-induced forgetting
d. Cognitive fade

2 **What does the brain do during sleep to help memory and focus?**

a. Stores every detail from the day
b. Doubles its activity
c. Deletes unimportant information
d. Regenerates new memories

3 **What happens in your brain when you're sleep-deprived?**

a. It shuts down temporarily
b. It forgets how to form memories
c. It begins devouring parts of itself
d. It enters "power-saving" mode

4 **What's true about multitasking and the brain?**

a. It enhances memory and focus
b. It allows better time management
c. Your brain switches between tasks, not multitasks
d. It's a skill that only some people have

5 **What happens each time you recall a memory?**

a. The memory strengthens and becomes more vivid
b. The memory remains exactly the same
c. Your brain rewrites them like a script
d. Your brain deletes similar memories

6 **Why is daydreaming good for your brain?**

a. It improves short-term memory
b. It raises your energy levels
c. It boosts creativity and problem-solving
d. It resets emotional balance

7 **What is the "rubber hand illusion"?**

a. A magician's trick for brain training
b. Your brain can adopt a fake limb as its own
c. A sensory disorder
d. A therapy for phantom limb pain

8 **What is neuroplasticity?**

a. A brain disease
b. A type of brain surgery
c. The brain's ability to change with experience
d. A tool for brain mapping

9 **How do gut bacteria influence the brain?**

a. They help prevent brain shrinkage
b. They affect your mood through the bloodstream
c. They increase memory
d. They improve attention span

10 **What do brain rhythms do?**

a. Help you sleep more deeply
b. Act like natural music regulating memory and motion
c. Create hallucinations
d. Form new neurons

Bonus Round – True or False (11–15)

11 **Circle True or False.**
Scientists can now control prosthetic limbs with brain waves.
True / False

12 Teen brains grow more connections over time to become more efficient. **True / False**

13 Scientists have mapped every single part of the human brain.
True / False

14 The brain contains "place cells" that help us navigate our environment. **True / False**

15 The brain stops developing after age 12. **True / False**

Answer Key

MULTIPLE CHOICE:
1. c 2. c 3. c 4. c 5. c
6. c 7. b 8. c 9. b 10. b

TRUE OR FALSE:
11. True 12. False 13. False 14. True 15. False

SCORING: HOW BRAINY ARE YOU?

Count how many questions you got right (out of 15):

13-15 correct: Brainiac Supreme - You're a neural ninja! The brainy elite. Science high-five!

10-12 correct: Clever Cortex - Solid work! You're sharp and soaking up science like a champ.

7-9 correct: Curious Cortex-in-Training - You've got potential! Give it another read and keep growing those synapses.

4-6 correct: Distracted Neuron - No worries—your brain was probably just multitasking (bad idea, remember?).

0-3 correct: Total Brain Fog - Hey, even Einstein had off days. Read it again with a good snack and a nap!

Heart to Heart: Cutting-Edge Science Facts That Will Make Yours Skip a Beat

Let's dive into the latest research that proves your heart has a lot more to say than just "thump-thump."

Your heart is a dynamo of energy, emotion, resilience and surprising science. From pollution's surprising impact to futuristic mini-hearts and cutting-edge risk calculators, these discoveries will give you a whole new appreciation for the ticker that keeps you going.

1

THE HEART'S "SUPERPOWER" FIELD

Your heart isn't just a pump; it generates the largest rhythmic electromagnetic field in your body, even stronger than your brain's. This field extends several feet around you and changes depending on your emotions: when you're happy, it's stronger, and when you're sad, it weakens. It's like having an invisible force that not only affects your own body but can also influence those around you. Imagine sending a "love wave" to your friends just by feeling good.

Source:
https://www.heartmath.org/research/science-of-the-heart/energetic-communication/

2

Your Heart Keeps Pumping When Disconnected From Your Brain

Your heart does not wait for your brain's permission to beat. It has its own built-in electrical system, a natural pacemaker, that keeps it pumping even if it's completely disconnected from the brain.

But here is where it gets even cooler. Today's doctors are using cutting-edge artificial intelligence to monitor and interpret your heart's electrical signals in real time. AI can spot subtle changes and predict problems before you feel a thing, making heart health checks faster, smarter, and more accurate than ever.

With wearable devices and smart monitors powered by AI, your heart's electric rhythm can be tracked around the clock, giving you and your doctor a high-tech window into your most vital organ.

Source:
https://www.health.harvard.edu/heart-health/artificial-intelligence-in-cardiology

3

Hearts That Don't Get Old?

Here's a heart-stopping discovery: Scientists are making real progress toward helping the heart stay young, even as the rest of the body ages.

Recent breakthroughs show that while the adult heart has a limited ability to regenerate on its own, researchers have now found ways to encourage heart muscle cells (cardiomyocytes) to multiply and repair damage. In 2025, scientists revealed a novel approach that stimulates these cells to proliferate, offering new hope that, one day, our hearts could keep regenerating and stay youthful far longer than we ever imagined.

Source:
https://www.sciencedaily.com/releases/2025/03/250307144708.htm

4

DOES THE HEART HAVE A MEMORY?

Some heart transplant recipients have reported developing new tastes, habits, or even personality traits that eerily match those of their donors. These include suddenly craving foods they never liked before or picking up new hobbies. Scientists are fascinated by these stories and are investigating whether something called "cellular memory" could be at play, where cells in the heart might somehow carry traces of the donor's preferences or experiences.

While this phenomenon is rare and still not fully understood, recent studies suggest that the heart's own neural network and its constant communication with the brain could play a role in shaping memory and personality. Researchers are exploring the idea that memories and traits might be transferred through the heart's cells, DNA, or even its unique electrical signals. It's a mystery that challenges what we know about identity.

Source:
https://pubmed.ncbi.nlm.nih.gov/38694651/

5

Stem Cells to the Rescue

If your heart gets damaged, scientists have figured out a way to use stem cells (special cells that can become anything) to fix it, like tiny heart mechanics.

In 2025, researchers reported a breakthrough: patches of heart muscle grown from stem cells were grafted onto a woman's failing heart, keeping her stable until she could get a transplant. These lab-grown patches stayed in place and formed new blood vessels, showing real promise for repairing damaged hearts. This could mean that, in the future, fixing heart damage might be as routine as treating a small wound.

Source:
https://www.nature.com/articles/d41586-025-00273-2

6

Your Heart Can Make Music

Your heart has its own unique rhythm that can actually be transformed into music.

Scientists have found that by recording the beat patterns of your heart, they can create a "heartbeat song" - a personal soundtrack that is as individual as your fingerprint. Recent studies show that music can influence your heart's rhythm, making it more regular and stable, and even promote relaxation and healing. Research published in 2025 found that listening to calming music can help regulate heart rhythms, suggesting a powerful connection between music, emotion, and heart health. This discovery is opening new doors for music therapy, where your own heartbeat could one day be part of your personalized path to relaxation and well-being.

Source:
https://publishing.aip.org/publications/latest-content/music-can-touch-the-heart-even-inside-the-womb/

7

Your Heart is the Ultimate Energy Powerhouse

Imagine your heart as a nonstop energy factory, working 24/7 without ever taking a break. Inside each heart cell are tiny power plants called mitochondria, cranking out the energy needed to keep your heart beating strong all day, every day. These little dynamos make up nearly half the volume of your heart cells and produce enough fuel to power your heart's relentless rhythm.

If you added up all the energy your heart generates over a lifetime, it could power a light bulb for years! Scientists are now discovering ways to keep these microscopic powerhouses in tip-top shape, helping your heart stay powerful and healthy as you age.

Next time your heart races, whether from excitement, exercise, or just life's hustle, remember it's your body's ultimate superhero, tirelessly fueling every beat and every move you make.

Source:
https://www.ahajournals.org/doi/10.1161/

CIRCRESAHA.124 https://www.sciencedaily.com/releases/2025/03/250320144617.htm.323800

8

Cool New Tech for Your Heart's Rhythm

If your heart beats too fast or too slow, doctors used to rely on heat or extreme cold to "fix" it. Now, a new breakthrough called pulsed field ablation (PFA) is changing the game. This technique uses high-energy electrical pulses-like a gentle zap-to precisely target and modify the heart tissue causing abnormal rhythms, while minimizing damage to surrounding healthy areas. It is a safer, more efficient way to nudge your heart back into its natural rhythm and is quickly becoming the preferred method for treating conditions like atrial fibrillation.

Source:
https://www.medtechdive.com/news/HRS-PFA-studies-Boston-Scientific-Abbott-Medtronic-JNJ/746659/

9

Your Heart's Secret "Little Brain"

Did you know your heart has its own "mini-brain"? Recent research has uncovered that the heart possesses an intrinsic cardiac nervous system (IcNS), a complex network of neurons capable of processing information independently from the brain. This discovery suggests that your heart can make certain decisions on its own, influencing both heart function and overall well-being. It is like having a smart co-pilot working tirelessly within you.

Source:
https://www.perplexity.ai/search/find-the-latest-open-source-fo-AO_jiG8GTJKhZPMpmMs5WQ

10

AI and Your Heart - The Ultimate Tag Team

Artificial intelligence (AI) is transforming heart health like never before. By analyzing not just your genetics but also your heart rhythms, medical images, and lifestyle data, AI can predict risks like heart failure or irregular heartbeats long before you notice any symptoms. It's like having a super-smart robot doctor working alongside your healthcare team, spotting problems early and tailoring treatments just for you. And the future is even more exciting-AI could soon become your personal heart coach, offering real-time advice to help you live a longer, healthier life.

Source:
https://medicine.yale.edu/news-article/new-ai-tool-identifies-risk-of-future-heart-failure/

11

Pollution: A Hidden Killer for Heart Health

Groundbreaking research from the Victor Chang Cardiac Research Institute reveals that pollution is now a bigger threat to global health than war, terrorism, and even major infectious diseases like malaria and HIV-combined.

Pollution, including dirty air, noise, light, and chemical exposure, is responsible for over seven million premature deaths every year, with more than half of these linked directly to heart and blood vessel diseases.

Shockingly, one in five cardiovascular deaths worldwide is now caused by air pollution alone. The Institute's experts warn that the impact of pollution on our hearts is greater than many people realize, making it one of the most urgent -and preventable - health threats of our time. Tackling pollution could save millions of lives and dramatically reduce the burden of heart disease across the globe.

Source:
https://www.victorchang.edu.au/news/pollutants-climate-change-cvd

12

Mothers of Twins Face a Hidden Heart Risk After Birth

With twin pregnancies becoming more common thanks to fertility treatments and women having babies later in life, this research is a wake-up call.

Women who give birth to twins are nearly twice as likely to be hospitalized for heart disease within their first year compared to those with single babies.

Carrying twins puts extra strain on the heart and if high blood pressure was part of the pregnancy story, that risk can skyrocket and make heart complications more than eight times as likely. While most moms of twins won't face serious heart problems, the months after birth are a crucial time to watch heart health closely. Your heart has been working overtime, and now it needs some extra care.

Source:
https://www.sciencedaily.com/releases/2025/02/250203142730.htm

13

New Cholesterol Pill Offers Hope for Those Who Need More Than Statins

Heart disease is still the leading cause of death in the U.S. and is a big problem in many other countries. A breakthrough new cholesterol-lowering pill called obicetrapib is shining a light of hope for those struggling to control their "bad" LDL cholesterol - especially those who cannot tolerate statins or need extra help beyond what statins provide.

Recent U.S. clinical trials show that obicetrapib, often combined with another medication called ezetimibe, can slash LDL cholesterol by nearly 50%, a game-changing boost for heart health.

This new pill offers a powerful alternative or add-on for those who don't get enough benefit from statins alone or experience side effects. This could help millions of people stay healthier and reduce their risk of heart attacks and strokes.

Source:
https://consultqd.clevelandclinic.org/obicetrapib-ezetimibe-combo-therapy-halves-ldl-cholesterol-levels-relative-to-placebo

14

New Heart Disease Risk Calculator Spots Hidden Dangers

Cardio-vascular diseases are responsible for an estimated 17.9 million deaths every year according to the World Health Organization.

Researchers at the University of Oxford have launched QR4, a next-generation heart disease risk calculator that is changing the game for predicting your 10-year risk of heart disease and stroke. Unlike older tools, QR4 looks at a much wider range of risk factors including certain cancers, chronic lung conditions, and even pregnancy complications like pre-eclampsia and postnatal depression. This means QR4 can spot high-risk patients that other calculators might miss, making it a powerful new ally in the fight against heart disease.

Developed using health data from over 16 million people, QR4 has been shown to outperform existing risk scores used in the US and Europe. With this breakthrough, doctors can now deliver more personalized and effective prevention strategies-potentially saving thousands of lives by catching hidden risks before they become serious problems.

Source:
https://pharmatimes.com/news/new-heart-disease-calculator-more-accurately-predicts-patients-risk-of-cardiovascular-diseases/

https://www.ox.ac.uk/news/2024-04-18-new-heart-disease-calculator-could-save-lives-identifying-high-risk-patients-missed

15

Innovative Palm-Cooling Devices Aid Athletes

Palm-cooling devices, known as Narwhals, are gaining popularity among athletes from football to swimming for their ability to bring on a rapid recovery. Cooling the palms (or soles of the feet) normalizes the heart rate more quickly. They can bounce back from heat stress more effectively and in less time.

Source:
https://www.wired.com/story/palm-cooling-devices

16

3D-Printed 'Mini Hearts' Offer Surgical Alternatives

Researchers at the Heart Research Institute (HRI) are pioneering the use of 3D-printed "mini-hearts" to create personalized heart tissue patches. HRI is a not-for-profit, internationally recognized medical research institute based in Sydney, Australia.

Made from a patient's own cells, these patches are designed to repair damaged heart tissue through minimally invasive keyhole surgery, offering a promising alternative to traditional open-heart surgery and potentially reducing the need for heart transplants.

HRI's focus on cutting-edge cardiovascular research aims to transform treatment options for heart failure patients in Australia and around the world as this technology advances toward clinical trials.

Source:
https://www.hri.org.au/our-research/research-groups/cardiovascular-regeneration

17

The Soaring Cost of Heart Disease

Here's a jaw-dropping projection: by 2050, the money the United States spends on heart disease and its risk factors-including high blood pressure, diabetes, and high cholesterol, is expected to triple. That means what cost about $400 billion in 2020 could skyrocket to over $1.3 trillion every year just to manage the risks. Add in the cost of actually treating heart attacks, strokes, and other heart conditions, and the total could reach a staggering $1.8 US trillion annually.

The population is getting older, and more people are developing conditions that put extra strain on the heart. Without stronger prevention and healthier habits, heart disease could become one of the biggest drains on the healthcare system and individual wallets.

Source:
https://www.heart.org/en/news/2024/06/04/heart-disease-and-stroke-could-affect-at-least-60-percent-of-adults-in-us-by-2050

18

Rethinking 'Bad' Cholesterol - Surprising New Findings

For decades, we've been told that high levels of LDL cholesterol, often called "bad" cholesterol, are a major culprit behind heart attacks and strokes. But new research is challenging this long-held belief.

A 2025 study found that lowering LDL cholesterol with medications like statins doesn't always lead to fewer heart attacks, strokes, or longer lives for everyone. This discovery is turning heads in the medical world and suggests that the link between LDL cholesterol and heart disease risk may not be as straightforward as once thought.

Source:
https://www.sciencedaily.com/releases/2025/04/250428222148.htm

19

Beyond Cholesterol: A Better Way to Predict Heart Risk

If cholesterol isn't the whole story, what is? Scientists are now discovering that counting the number of "bad cholesterol" particles in your blood, something called apoB, could be a much better way to predict who's really at risk for heart disease.

Researchers say that for about one in twelve patients, standard cholesterol tests may underestimate heart disease risk. This is important to consider, since 20 to 40 percent of all first-time occurrences of cardio-vascular diseases are fatal. They believe that switching to apoB testing will improve that accuracy and potentially save lives.

Source:
https://www.sciencedaily.com/releases/2025/04/250428222148.htm

20

Exciting Advances in Cardiovascular Disease (CVD) Treatment

Cardiovascular care is entering a new era, with breakthrough technologies offering hope and better outcomes for millions. Today, patients can benefit from minimally invasive heart procedures such as:

- Robotic-assisted and keyhole surgeries that restore blood flow or repair valves without the trauma of traditional open-heart surgery. These techniques mean smaller scars, faster recovery, and less time in the hospital.
- A new generation of anti-obesity drugs is making headlines not just for helping with weight loss, but for slashing the risk of heart attacks and strokes by up to 20% in people with obesity or diabetes.
- AI-powered tools are now helping doctors detect heart problems earlier and more accurately.
- Gene-editing therapies like CRISPR are opening the door to treating inherited heart diseases that were once considered untreatable.

With these innovations, the future of heart health is brighter than ever, offering safer, more personalized, and more effective treatments for patients everywhere.

Source:
https://pmc.ncbi.nlm.nih.gov/articles/PMC11763404/#main-content

Latest Science Discoveries with Quirky Twists

Step into a world where science gets stranger, smarter, and more surprising with every new discovery.

This collection of the latest research and oddball breakthroughs will have you rethinking what's possible from birds with a built-in quantum compass to brainy slime molds and golden-tongued mummies.

1

Quantum Compass - How Birds Use Physics to Fly Home

It sounds like science fiction, but it's real: birds use quantum mechanics to navigate thousands of miles during migration. Researchers have discovered that migratory birds, like robins, can actually detect Earth's magnetic field using a light-sensitive protein in their eyes called cryptochrome.

Here's where it gets wild—when light hits this protein, it triggers a quantum reaction that alters the spin of electrons, helping the bird "see" magnetic fields as visual patterns. This internal compass allows them to track direction and location with remarkable precision, even on cloudy days or across vast oceans.

This ability is powered by a quantum process called radical pair-based magnetoreception, and scientists have used quantum computers to simulate exactly how it works. The birds don't need GPS, a map, or even the stars—just the Earth's magnetic field and some physics most of us barely understand.

Next time you see a bird on its migration journey, remember it is navigating with built-in quantum tech that modern science is only just beginning to decode.

Source:
https://modernsciences.org/new-study-reveals-genetic-basis-of-migratory-birds-navigation-abilities/

2

YOUR GUT BUGS MAY BE CONTROLLING YOUR MOOD

Ever had a "gut feeling", feel anxious for no clear reason, or suddenly upbeat after a good meal? Turns out your gut bacteria may be calling the shots on more than just digestion.

Inside your digestive system live trillions of microbes—collectively known as your gut microbiome—and they're not just helping with digestion. Scientists have discovered these microbes are in constant communication with your brain through what's called the gut-brain axis.

This isn't fringe science—it's cutting-edge research. Certain gut bacteria can actually produce neurotransmitters like serotonin, dopamine, and GABA—the same brain chemicals that affect your mood, focus, and stress response. These signals travel through nerves, hormones, and even the immune system, shaping how you feel emotionally and So if your mood suddenly dips or your sleep patterns feel off, your gut might be involved. Researchers are now linking imbalances in gut bacteria to conditions like anxiety, depression, and even sleep disorders. It's a growing area of science suggesting that taking care of your gut could be one key to supporting your mental well-being—along with your digestion.

Source:
https://www.nature.com/articles/s41392-024-01743-1

3

Octopuses Can Reprogram Their Genes

Octopuses aren't just shapeshifters on the outside—they can reprogram themselves on the inside at the genetic level too.

New research reveals that these brainy sea creatures have a rare ability to rapidly edit their own RNA, especially in their nervous systems. This lets them fine-tune how their bodies function without changing their DNA.

In one study, scientists cooled the water in octopus tanks and saw a dramatic spike in RNA editing—at more than 13,000 sites in the nervous system. These edits directly affected proteins crucial for brain and nerve function, helping the octopuses quickly adjust to the colder environment.

This remarkable genetic flexibility is one reason why octopuses can survive—and thrive—in such wildly different ocean conditions.

Source:
https://news.uchicago.edu/story/octopuses-other-cephalopods-can-adjust-cold-editing-their-rna

4

WATER'S MIND-BENDING MAGIC

Water seems simple, right? H2O, drink it, splash it, swim in it. New research shows that under extreme conditions, like intense cold or pressure, water can act like two different liquids at once. It shifts between a low-density and high-density form, as if it's living a double life.

This bizarre behavior may explain why water breaks so many scientific "rules" such as why ice floats, or why water expands when it freezes. Researchers think this hidden transformation is the key to understanding water's many mysteries. So the next time you pour a glass, remember that water doesn't just flow - it bends the rules of reality.

Source:
https://www.sciencedaily.com/releases/2024/11/241114161415.htm

5

Electric Fish – the Underwater Spy Game

Electric fish are nature's underwater secret agents. They generate electric fields not only to navigate murky waters and communicate with each other but also to eavesdrop on the electric signals of other fish.

Some species can even "cloak" their own electric signals to avoid detection by predators, while others pick up on rivals' or prey's electric cues to outsmart them. This underwater spy game involves sophisticated electric "codes" and stealth tactics, making these fish masters of aquatic espionage.

Source:
https://www.frontiersin.org/articles/10.3389/fevo.2019.00264/full

6

Talking Mummies with Golden Tongues

Recent discoveries have made ancient Egypt even more dazzling. In early 2025, archaeologists unearthed 13 mummies at Oxyrhynchus with golden tongues-thin gold amulets placed in their mouths over 2,000 years ago.

The ancient Egyptians believed these golden tongues would let the dead speak in the afterlife, especially to gods like Osiris. These burials weren't just about tongues; some mummies had golden fingernails, and the tombs were filled with ritual texts, vibrant paintings, and amulets. It's ancient bling with a purpose: giving the deceased a divine voice and a little extra sparkle for their journey beyond.

Source:
https://www.popularmechanics.com/science/archaeology/a63412049/golden-mummy-tongues/

7

Your Dog's Ancestry Is a Global Tale

Think you know where dogs came from? The story is more fascinating than ever.

Some earlier studies suggested dogs might have been domesticated twice-once in Europe and once in Asia. However, recent genetic research points to a single domestication event, likely somewhere in Eurasia, between 20,000 and 40,000 years ago.

After that, dogs spread with humans across continents, diverging into eastern and western lineages as they traveled. So, your furry friend's family tree is still a globe-trotting adventure, shaped by ancient migrations, mixing, and the deep bond between dogs and people.

Nature Communications, "Ancient European dog genomes reveal continuity since the Early Neolithic"

Source:
https://www.nature.com/articles/ncomms16082

8

BRAINY SLIME MOLD SOLVES MAZES

Slime molds have stunned scientists by solving mazes – without a brain.

Despite having no brain or nervous system, these single-celled organisms can solve complex problems, such as finding the shortest path through a maze to reach food.

Even more astonishing, a 2021 study found that slime molds can "remember" past food locations by weaving memories directly into their body structure, allowing them to make smarter decisions in the future. This means slime molds don't just solve puzzles, they use a form of primitive memory to optimize their search, outsmarting expectations for brainless life.

Imagine something without a brain not only beating you at a maze but also learning from its past moves - a true natural genius hiding in plain sight.

Source:
https://phys.org/news/2021-02-single-celled-slime-mold-nervous-food.html

9

Tattoo Ink Travels to Your Lymph Nodes

Tattoos aren't just skin-deep. Scientists recently found that tattoo ink travels to your lymph nodes.

Within hours of getting a tattoo, ink particles are transported via the lymphatic system and can be found in nearby and even distant lymph nodes. These particles are mainly captured by immune cells called macrophages and can remain in the lymph nodes for months or even years.

Some studies have also observed that this accumulation can trigger both acute and long-lasting inflammation in the lymph nodes, though the long-term health effects are still being investigated. While there's ongoing research into whether this ink migration could be linked to increased risks of certain cancers, no definitive harmful effects have been established for most people.

So, that dragon tattoo on your arm? Part of its ink may have hitchhiked to your lymph nodes, making tattoos even more permanent than we once thought.

Source:
https://www.sciencedaily.com/releases/2025/03/250303141854.htm

10

The Sneaky Life of Carnivorous Plants

You may have heard of Venus flytraps, but other plants, like the Cape sundew, don't need any flashy traps to catch their prey. These sticky, hungry plants capture insects and digest them slowly without any sudden snaps - just a sticky, sneaky surprise for their victims.

New research shows one of nature's slow-motion insect trappers is far more adaptable than we once believed. Sundews can actually dial up or down their meat-eating behavior depending on their environment. When nutrients are low but sunlight is strong, they grow more sticky tentacles and go full predator mode, catching extra insects to make up the difference. In richer soil or low light, they chill out and conserve energy.

This flexible strategy makes sundews expert survivors in harsh, changing landscapes like bogs and peatlands. While they may not snap shut like Venus flytraps, they're quietly brilliant at adjusting their tactics and proof that, in the world of carnivorous plants, brains (and goo) can beat speed.

Source:
https://phys.org/news/2025-01-sundews-secret-survival-reveals-carnivorous.html

Nature's Amazing Hidden Talents: Fresh Discoveries That Redefine the Wild

Here you will see the natural world in a whole new light! This collection of recent research facts reveal the astonishing superpowers, secret languages, and quirky survival tricks found in creatures and plants all around us.

From butterflies with ultraviolet vision and cacti with natural sunscreen to singing ice, "immortal" jellyfish, and cats whose purrs may heal, these discoveries prove that nature's creativity knows no bounds. Whether you're a science buff, a nature lover, or just someone who loves a good surprise, these facts are sure to inspire awe, delight, and maybe even spark your own next big idea.

1

Butterflies and Their Super Sight

Butterflies have incredible vision that goes beyond human limits. Unlike us, they see ultraviolet light, which makes flowers glow like neon signs, guiding them to tasty nectar. Scientists have discovered that butterfly eyesight is adapted to detect these UV patterns on flowers, a skill humans only developed after creating UV technology in the lab.

In fact, some butterfly species possess up to 15 types of photoreceptor cells, compared to the three found in human eyes, enhancing their capacity to detect a broader range of light wavelengths. Who's the real tech genius here?

Source:
https://kids.frontiersin.org/articles/10.3389/frym.2017.00070

2

The Cactus and Its Built-In Sunblock

In a twist on nature's sunscreen, desert cacti produce a chemical layer on their skin that protects them from the harsh desert sun. This "sunscreen" helps them survive blazing heat without drying out. Botanists have recently examined this natural layer, inspiring scientists to consider new eco-friendly sunscreens for humans.

Source:
https://www.researchgate.net/publication/336284826_Antioxidants_and_cactus_extract_enhances_the_sun_protection_factor_of_octocrylene_and_oxybenzone_solution

3

Mystery of the Starfish's Invisible Brain

The starfish has no centralized brain, yet it can still hunt, move, and react to danger. Scientists in 2024 used neural mapping to study its nerve ring—a circular "brain" that sends signals to each limb independently. Starfish prove that thinking doesn't always need a brain as we know it.

Source:
https://www.nature.com/articles/s41598-023-30425-1

4

Spiders and the Secrets of Their Webs

Spider silk is five times stronger than steel of the same thickness. Researchers are studying how spiders can produce such strong threads in hopes of creating similar materials for surgery or engineering. We might have spider webs to thank for future tech in medicine and construction.

Source:
https://www.science.org/content/article/spider-silk-five-times-stronger-steel-now-scientists-know-why

5

Music and Plants: A Growing Relationship

Plants may actually "hear" sounds around them. Recent research suggests that certain plants respond to frequencies produced by insects and other plants. Plants exposed to specific musical tones grew healthier and taller, sparking ideas about how sounds could be used in farming to boost crop growth.

Source:
https://gardenculturemagazine.com/good-growing-tunes-how-plants-respond-to-music/

https://justagriculture.in/wp-content/uploads/2024/12/36.-Plant-Acoustics-How-Plants-Sense-and-Respond-to-Sounds.pdf

6

Ancient Teeth Hold the Key to Diet Mysteries

Archaeologists studying ancient tooth fossils recently found traces of starchy plants, showing that humans ate carbs thousands of years earlier than once thought. This discovery challenges old theories, giving insight into early diets and how they shaped our health and evolution.

Source:
https://www.nature.com/articles/s41415-022-5266-7

7

Shocking Chatters: How Electric Eels "Talk" Underwater

Electric eels don't just zap to hunt—they also use gentle electric pulses to "talk" to each other? These low-voltage zaps are like secret underwater Morse code, helping eels navigate dark waters, warn rivals, and find mates. Scientists are so amazed by this natural electric language that they're designing new underwater communication tech inspired by these slippery shockers. Next time you think of electric eels, remember—they're not just hunters, they're also expert communicators lighting up the river with their electric conversations!

Source:
https://nationalzoo.si.edu/animals/electric-eel

8

Jellyfish Genes and the Secret to Eternal Youth

The "immortal jellyfish" can reverse its aging process, turning back into a juvenile state under certain conditions. Scientists studying this jellyfish's genes are trying to unlock secrets that could one day slow human aging.

Source:
https://www.smithsonianmag.com/smart-news/immortal-jellyfish-could-spur-discoveries-about-human-aging-180980702/

9

Desert Frogs Have Built-In "Water Bottles"

To survive harsh droughts, desert frogs absorb and store water directly in their bodies, making their own built-in reservoirs. When water is scarce, they rely on this store, a trick that fascinates researchers looking at hydration methods in extreme environments.

Source:
https://www.australiangeographic.com.au/topics/wildlife/2018/05/meet-australias-desert-dwelling-frogs/

10

Octopuses Rewrite Their Own DNA

Octopuses can "edit" their RNA, allowing them to adapt their nervous systems to new environments. Imagine if we could edit ourselves on the spot to be better swimmers or thinkers. Scientists are hoping to learn more about these adaptations and apply them to human health.

Source:
https://www.sciencedaily.com/releases/2023/06/230608120915.htm

11

Plants May "Scream" When Hurt

Plants might not have voices, but when they're stressed-like being thirsty or getting a cut-they actually "scream" in a way. Using special ultrasonic microphones, scientists discovered that tomato and tobacco plants emit tiny popping sounds, too high-pitched for human ears, but possibly audible to some animals nearby. It's like plants have their own secret language of distress, quietly signaling trouble before we even see the damage.

Source:
https://www.smithsonianmag.com/smart-news/plants-make-noises-when-stressed-study-finds-180981920/

12

Sharks' Sixth Sense: Electric Fields

Sharks have special organs called ampullae of Lorenzini that detect electric fields generated by animals' movements, helping them locate prey even in total darkness. This "sixth sense" is a skill that researchers are studying to develop new underwater navigation tech.

Source:
https://en.wikipedia.org/wiki/Ampullae_of_Lorenzini

13

Insects and the Power of "Super-Genes"

Certain insects like ants and bees have "supergenes" that determine complex behaviors, from building elaborate homes to forming societies. Scientists are exploring how these genes could help in studying social behavior in other animals, including humans.

Source:
https://blog.myrmecologicalnews.org/2023/01/11/the-supergene-wave-is-well-underway/

14

Birds Use Quantum Physics for Migration

Migratory birds use quantum mechanics to "see" Earth's magnetic field, helping them navigate long distances. Their eyes contain proteins that allow this navigation method, a discovery that's inspiring new tech in GPS-free navigation.

Source:
https://www.sciencenews.org/article/quantum-mechanics-compass-songbird-physics

15

Mushrooms and the Wood Wide Web

Mushrooms connect with trees through underground networks, sharing nutrients and signals. This "Wood Wide Web" is a model for studying complex systems in ecology, inspiring scientists to look at ways of enhancing crop resilience.

Source:
https://www.uni-bayreuth.de/en/press-release/mycoheterotrophic-plants

https://biosciences.lbl.gov/2024/04/24/an-inside-look-at-how-plants-and-mycorrhizal-fungi-cooperate/

16

Naked Mole-Rats are Age-Defiers

Naked mole-rats resist cancer and live long lives, despite their odd looks. Studying them might reveal new anti-aging secrets for humans, as they possess unique genes that slow down aging and enhance cell repair.

Source:
https://www.aginganddisease.org/EN/10.14336/AD.2024.0109

https://pmc.ncbi.nlm.nih.gov/articles/PMC11745443/

17

Cockroach "Milk" as a Superfood

Researchers found that a species of cockroach produces milk with high protein and energy content. This "milk" could be a sustainable food source in the future, though most people aren't quite ready for bug-based snacks.

Source:
https://pmc.ncbi.nlm.nih.gov/articles/PMC6123606/

18

Electric Fish Communicate in Morse Code

Electric fish create unique pulses to "talk" to each other underwater. Each species has its own pulse pattern, much like Morse code, sparking ideas for new ways of coding information for human tech.

Source:
https://en.wikipedia.org/wiki/Electric_fish

19

Ancient Bacteria Found Deep Underground

Scientists found bacteria that have survived for millions of years in isolation, deep within Earth's crust. These resilient bacteria could teach us about life in extreme environments and may even help us search for extraterrestrial life.

Source:
https://www.earth.com/news/microbes-found-alive-sealed-inside-rock-after-2-billion-years/

20

Robots Help Animals "Chat" Brain-to-Brain

In 2025, Chinese scientists built a tiny flexible robot that can gently implant ultra-thin wires into the brains of rats and monkeys. These electrodes act like translators, picking up brain signals and sending them between animals or machines. This lets animals share thoughts and decisions directly—like a secret brain-to-brain chat—opening new ways to understand and connect animal minds.

Source:
https://en.people.cn/n3/2025/0430/c90000-20309659.html

21

How Bees Inspire Smarter Tech

Bees' incredible ability to zoom safely through cluttered spaces is inspiring new tech. In 2025, researchers showed how bees use special vision tricks to avoid collisions at high speed. Engineers are now using these insights to design drones that can fly through forests or cities without crashing. Studying how bees solve puzzles also helps make AI and robots smarter and more adaptable.

Source:
https://phys.org/news/2025-05-animal-cognition-reveals-treasure-trove.html

22

Chameleons Have Natural "Solar Panels" in Their Skin

Chameleons aren't just masters of camouflage; they have microscopic "solar cells" in their skin that reflect light and help regulate body temperature. Scientists think these cells act like tiny mirrors to scatter light, keeping chameleons cool even in the hottest environments. Some researchers are now exploring how this light-bouncing technique could inspire energy-efficient materials for our homes.

Source:
https://www.nature.com/articles/s41598-018-19498-5

23

Whale Songs Are Getting Deeper

Whale songs are like nature's greatest hits - constantly evolving. Recent research shows that blue, fin, and humpback whales are not only singing deeper and more complex tunes over time but also tweaking their melodies to match changes in their ocean environment. It is possible that the lower pitch travels longer distances deeper in the ocean, as if these ocean giants are remixing their songs to cut through murky waters and noisy seas, helping them to stay connected.

Source:
https://phys.org/news/2025-02-eavesdropping-whale-songs-discoveries-ecology.html

https://www.sciencedaily.com/releases/2021/09/210901113737.htm

24

Ancient Egypt's Blue Dye Was the World's First Fluorescent Material

The bright blue used in ancient Egyptian art and hieroglyphs isn't just eye-catching; it's also the earliest example of a man-made material that fluoresces. Scientists now believe the dye could have had special ceremonial significance, glowing under certain light conditions. Egypt's brilliant blue may even inspire new biomedical technology for tagging cells in medical research.

Source:
https://www.artinsociety.com/egyptian-blue-the-colour-of-technology.html

25

Birds Inspire New Navigation Technologies

Birds like robins and swans have proteins in their retinas that allow them to "see" Earth's magnetic field, guiding them on their long migrations. These molecules create a visual map based on the magnetic field that helps birds stay on course, even without landmarks or GPS. Scientists are studying this magnetic vision to develop new navigation technologies.

Source:
https://phys.org/news/2024-04-evolution-optimized-magnetic-sensor-birds.html#google_vignette

26

SUNSCREEN FOR PLANTS? IT'S REAL - AND INSPIRING NEW SKINCARE FOR HUMANS

Some desert plants have their own built-in sunscreen. To survive intense sunlight and harmful UV rays, these hardy plants produce special natural chemicals like salicylic esters that form a protective shield over their leaves. Scientists are fascinated by these plant-made UV blockers and are now exploring how to harness them to create eco-friendly, biodegradable sunscreens for humans. Imagine slathering on sunscreen inspired directly by nature's own SPF-a perfect blend of science and sustainability!

Source:
https://pmc.ncbi.nlm.nih.gov/articles/PMC10272908

27

THE ARCTIC OCEAN IS HOME TO "SINGING" ICE

A 2023 study demonstrates how melting and refreezing sea ice emits distinctive acoustic signals - often described as "singing" or "haunting" sounds-that can be detected underwater. These icy melodies are now being recorded by advanced sonar and passive acoustic monitoring systems, providing valuable data on ice thickness and the effects of climate change in the Arctic.

Source:
https://agupubs.onlinelibrary.wiley.com/doi/10.1029/2023GL104578

28

Cats' Purring Could Be Nature's Healing Frequency

While a cat's purr is often a sign of contentment, scientists have discovered that its frequency-typically between 25 and 150 Hz-may actually promote healing. These vibrations align with frequencies known to stimulate bone growth (25-50 Hz) and tissue repair (around 100 Hz). Research suggests that purring can reduce inflammation, encourage bone regeneration, and even alleviate pain. Beyond physical healing, the soothing sound of purring can lower stress hormones and promote emotional well-being in humans. This remarkable natural therapy is inspiring new approaches to sound-based healing.

Source:
https://www.bbc.com/future/article/20180724-the-complicated-truth-about-a-cats-purr

29

Starfish Don't Have Blood

Instead of blood, starfish use seawater to pump nutrients through their bodies, thanks to a unique water vascular system. This seawater "circulation" system has intrigued scientists studying how animals can adapt to extreme environments, and it might offer new ways to think about survival in space or underwater habitats.

Source:
https://animaldiversity.org/accounts/Asteroidea/

30

Tardigrades Can Survive Space

These ambling, eight-legged microscopic "bears of the moss" are cute and all but indestructible. Tardigrades can survive the vacuum of space, extreme temperatures like deserts, being frozen in ice and even radiation. Researchers are studying their unique genes for insights into extreme survival, hoping to apply some of their tricks to human cells.

Source:
https://www.americanscientist.org/article/tardigrades

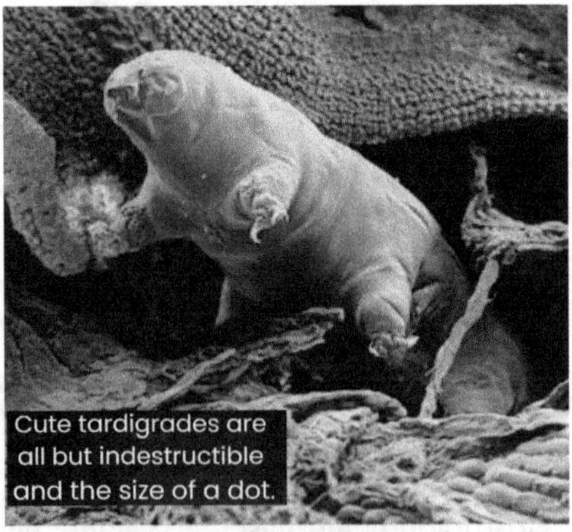

Cute tardigrades are all but indestructible and the size of a dot.

Credit: American Scientist. Eye of Science/Photo Researchers

31

Dogs Are Experts at Reading Human Emotions

Recent studies show that dogs can accurately distinguish between their owners' genuine emotions, such as happiness and sadness, and adjust their behavior accordingly. While scientists have not confirmed the presence of "empathy neurons" or mirror neurons in dogs, research suggests that dogs' ability to read and respond to human feelings is likely the result of specialized brain processing and a long history of close companionship with humans.

Source:
https://academic.oup.com/scan/article/19/1/nsae082/7884269

32

Jellyfish Are Earth's Oldest "Aliens"

Jellyfish have been around for more than 500 million years, making them older than dinosaurs and most other life forms on Earth. Their simplistic, alien-like design has hardly changed, making scientists wonder if these "living fossils" might teach us about life in extreme environments—maybe even on other planets.

Source:
https://www.smithsonianmag.com/science-nature/these-508-million-year-old-fossils-may-be-earths-oldest-swimming-jellyfish-180982639/

33

Frogs in Alaska – Frozen Alive

In one of nature's strangest survival tricks, some frogs in Alaska survive winter by freezing nearly solid. Their hearts stop, and their bodies turn icy, but they thaw back to life in spring. Studying this extreme hibernation may help scientists find ways to protect human organs during surgeries.

Source:
https://www.nps.gov/kova/learn/nature/wood-frog.htm

34

Sloths Grow Algae in Their Fur for Extra Nutrients

Sloths are so slow-moving that they allow algae to grow on their fur, giving them a greenish tint and a small snack. This relationship helps sloths blend in with their surroundings while also providing nutrients. Recent studies reveal that the algae in sloth fur may live in a lichen-like partnership with fungi, creating a complex micro-ecosystem. Additionally, the fur harbors antibiotic-producing bacteria that help protect sloths from infections, making their unique fur habitat both a source of nutrition and natural defense. Researchers are studying these interactions to better understand sloth health and explore new models for ecological camouflage and potential medical applications.

Source:
https://pmc.ncbi.nlm.nih.gov/articles/PMC10197869/

35

Parrots Name Their Babies.

Certain species of parrots use specific sounds as "names" for their chicks, which they continue to use even after their chicks are grown. For example, Green-rumped Parrotlets learn unique vocal signatures from their parents that function much like names, allowing family members to recognize and address each other individually. This remarkable ability is one of the first clear examples of naming in the animal kingdom and offers scientists valuable insights into the origins of language and complex social behavior in animals.

Source:
https://birdnote.org/podcasts/birdnote-daily/do-parrots-name-their-chicks

36

Octopus Arms Have "Minds" of Their Own

Octopuses' arms have a high concentration of neurons, allowing them to operate somewhat independently of their brain. This is like each arm having its own "mini-brain," capable of moving, exploring, and even solving puzzles without direct commands from the octopus's head.

Source:
https://www.cell.com/current-biology/fulltext/S0960-9822(22)01763-8

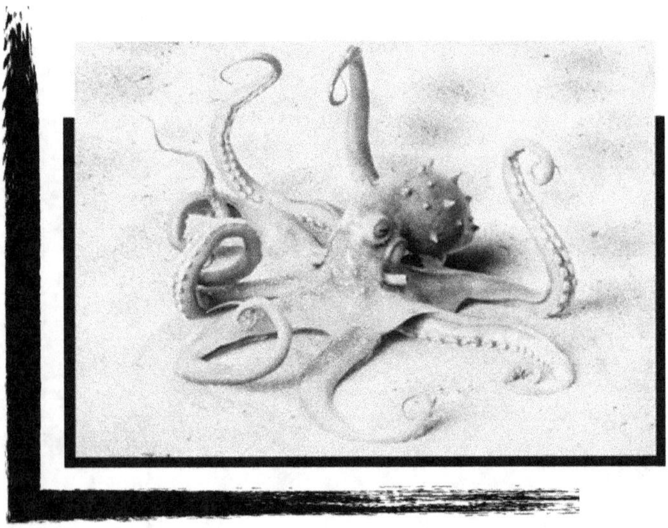

37

Some Bats Use "Superfast" Muscles for Screeching

Some bats use "superfast" muscles for screeching. The Mexican free-tailed bat has the fastest-moving muscles of any mammal, allowing it to produce rapid-fire ultrasonic calls for echolocation. These specialized muscles can contract up to 100 times faster than typical muscles, enabling bats to quickly adjust the pitch of their screeches and expertly navigate in complete darkness. Scientists believe that understanding these superfast muscles could inspire new advances in hearing aid technology, making devices more responsive to rapid sound changes.

Source:
https://www.sciencedaily.com/releases/2011/09/110929152102.htm

38

Termite Mounds Create "Natural Air Conditioning"

New research demonstrates that termite mounds use a dense, lattice-like network of tunnels and vents to harness wind and create interior turbulence, which powers ventilation and maintains stable temperature and humidity-even in extreme heat. Scientists are now using 3D-printed models of these structures to develop energy-efficient, fanless air conditioning systems for buildings, offering a blueprint for sustainable architecture that reduces reliance on traditional, energy-intensive cooling.

Source:
https://www.sustainabilitymatters.net.au/content/sustainability/article/sustainable-air-conditioning-inspired-by-termites-674121038

39

Orangutans Can Make "Umbrellas"

Wild orangutans have been observed using large leaves as "umbrellas" to shield themselves from rain, showcasing advanced problem-solving and tool use. Recent research also reveals that orangutans construct leafy roofs over their nests and even use plants for self-care, such as applying medicinal leaves to wounds. These clever behaviors highlight orangutans' remarkable adaptability and their ability to use tools and natural materials in ways that sometimes mirror human ingenuity.

Source:
https://www.science.org/content/article/orangutan-plays-doctor-heals-himself

40

Crocodiles Are Surprisingly Sensitive to Touch

Despite their tough, armored skin, crocodiles have incredibly sensitive snouts-more sensitive than human fingertips. Their snouts are covered with tiny sensory organs called integumentary sensory organs (ISOs) that detect minute changes in water pressure and vibrations, helping them locate prey even in complete darkness. This exceptional touch sensitivity rivals that of some primates and represents a unique sensory adaptation among reptiles, enhancing their hunting abilities and environmental awareness.

Source:
https://www.nationalgeographic.com/animals/article/121108-nile-crocodile-duncan-leitch-science-human-sensitive-touch

(podcast)
https://daffysroundtable.captivate.fm/episode/keeping-crocodiles-in-captivity-with-chris-dieter-of-crocodile-encounter

Bonus Fact

1

THE COSMIC MICROWAVE BACKGROUND

Did you know that the Cosmic Microwave Background (CMB) radiation is a remnant from the Big Bang? This faint glow permeates the cosmos and gives us a snapshot of what the universe looked like when it was just 380,000 years old. Scientists study this radiation to unlock secrets about the universe's formation and evolution.

In 2025, researchers released the clearest high-resolution images yet of this ancient radiation, captured by the Atacama Cosmology Telescope (ACT). They show the first steps towards making the earliest stars and galaxies.

Source:

https://phys.org/news/2025-03-clearest-images-year-universe-reveal.html

https://www.cfa.harvard.edu/research/topic/cosmic-microwave-background

https://www.scienceabc.com/nature/universe/what-is-the-cosmic-microwave-background.html

https://www.esa.int/Science_Exploration/Space_Science/Cosmic_Microwave_Background_CMB_radiation

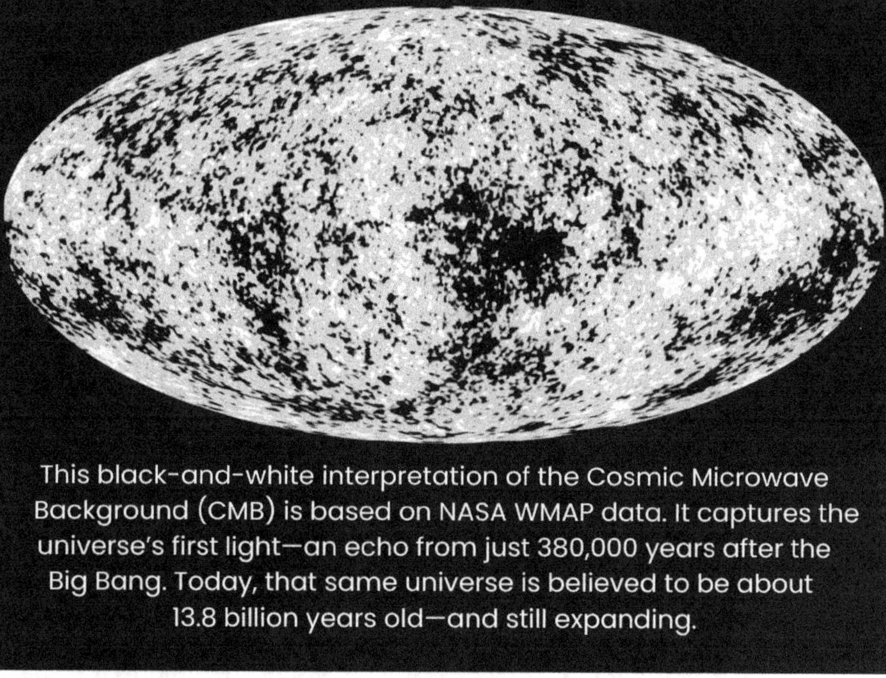

This black-and-white interpretation of the Cosmic Microwave Background (CMB) is based on NASA WMAP data. It captures the universe's first light—an echo from just 380,000 years after the Big Bang. Today, that same universe is believed to be about 13.8 billion years old—and still expanding.

Image credit:
NASA / WMAP Science Team

Ancient Hacks, Modern Proof

Turns out, ancient rituals weren't just folklore after all. Modern science is now confirming what many ancient cultures already figured out - sometimes thousands of years ago.

These fun facts show how the old ways are making a surprising comeback, with real evidence to back them up.

1

MEDITATION AND THE MIND

Ancient Insight: For thousands of years, cultures in India and China have used meditation to calm the mind and heal the body. Yogis and monks believed that by focusing on breathing and quieting the thoughts, people could reach a higher state of awareness and improve mental health.

Modern Science: Recent breakthroughs using advanced brain imaging and intracranial EEG technology reveal that meditation doesn't just help you relax, it actually rewires your brain in profound ways. Even beginners experience remarkable changes deep inside the amygdala and hippocampus, the brain's emotional and memory centers. For long-term practitioners, these changes become even more dramatic, with enhanced brainwave patterns that supercharge focus, emotional resilience, and stress relief.

It's like giving your brain a full reset button, unlocking your mind's incredible potential for calm, clarity, and well-being.

Source:
https://www.sciencedaily.com/releases/2025/02/250204132018.htm

2

Herbal Medicine vs. Pharmaceuticals

Ancient Insight: Indigenous peoples worldwide, from the Amazon to Africa, have used plants to treat illnesses. They often made teas, poultices, or extracts from herbs like turmeric, ginger, and willow bark for to relieve pain relief, inflammation and fever. The ancient Sumerians, circa 2000 BC, recorded the use of willow leaves to treat inflammation. The Ebers Papyrus, circa 1500 BC, is one of the most important medical documents of ancient Egypt and references the use of willow leaves as a general-purpose pain reliever.

Hippocrates himself advocated the use of willow bark tea to reduce fever and alleviate pain. Its active compound, salicin, is converted into salicylic acid in the body, providing effects similar to aspirin but with milder and longer-lasting benefits.

Modern Science: Today, scientists confirm that many of these plants have powerful medicinal properties. For instance, aspirin, one of the world's most widely used drugs, was derived from the salicylic acid found in willow bark. In turmeric the active component is curcumin which has been shown to reduce inflammation and is recognized for its anti-inflammatory and antioxidant properties. Studies have highlighted its potential to alleviate conditions like arthritis and support overall health when consumed regularly.

Source:
https://www.mountsinai.org/health-library/herb/willow-bark

https://pmc.ncbi.nlm.nih.gov/articles/PMC8572027

3

GROUNDING: FEET ON THE EARTH

Ancient Insight: Native American and many Eastern cultures believed that walking barefoot on the earth, known as grounding or earthing, was essential for balancing energy and maintaining good health.

Modern Science: Recent studies have shown that walking barefoot on natural surfaces like beach sand and grass can reduce inflammation, improve sleep, and boost mood. This happens because the Earth's electrons help neutralize free radicals in our body, providing antioxidant effects.

Source:
https://www.ncbi.nlm.nih.gov/pmc/articles/PMC3265077/

4

Breathing Techniques and Stress

Ancient Insight: Ancient yogis in India practiced Pranayama, a set of breathing techniques designed to control life force energy. They believed that different breathing patterns could calm the mind, energize the body, and even heal diseases.

Modern Science: Research now shows that deep breathing exercises directly affect the nervous system, reducing cortisol (the stress hormone) and lowering heart rate. Controlled breathing stimulates the parasympathetic nervous system, which calms the body and mind.

Source:
https://pubmed.ncbi.nlm.nih.gov/31073353/

5

SOUND THERAPY MAKING WAVES IN MODERN MEDICINE

Ancient Insight: Tibetan monks and Native American shamans used sound healing with instruments like gongs, bowls, and chants, believing that specific vibrations could heal the body by restoring harmony to its energy fields. In Egypt, priests used chanting and ritual instruments for healing. In India, practices like Nada Yoga and mantras aimed to harmonize mind and body. In China, sound and music were used to balance vital energy (Qi) and, in Greece, philosophers such as Pythagoras believed in the therapeutic effects of musical vibrations to restore health and emotional balance.

Modern Science: The science is clear: the right sounds, at the right frequencies, are powerful tools for mental and emotional well-being.

Recent studies reveal that sound interventions including music, nature sounds, and even specific vocal tones, can lower stress by calming the brain's stress response systems. Excitingly, brainwave entrainment techniques using binaural beats and isochronic tones have been shown to boost alpha brainwaves, helping to reduce anxiety, depression, and insomnia while improving sleep quality.

Even more groundbreaking, non-invasive sound wave therapies are now being used to target deep brain regions, offering real relief from depression, anxiety, and PTSD.

Source:
https://pubmed.ncbi.nlm.nih.gov/39699823/

6

THE MIND-BODY CONNECTION

Ancient Insight: In Chinese medicine and Ayurvedic traditions, amongst others, it is believed that emotions are stored in the body and can manifest as physical illness if not dealt with properly. For example, fear is said to affect the kidneys, while anger harms the liver.

Modern Science: Modern medicine is increasingly validating what ancient systems like Ayurveda and Chinese medicine have long taught: our emotions and mental states can profoundly influence physical health. While traditional practices described specific links to particular organs, contemporary research confirms that chronic stress, unresolved emotions, and poor emotion regulation are associated with a wide range of physical illnesses, including heart disease and immune dysfunction. Today, doctors recognize that caring for both mind and body is essential for true well-being, blending ancient wisdom with scientific evidence to support holistic health.

Source:

https://www.unsw.edu.au/newsroom/news/2025/05/new-chronic-pain-therapy-retrains-brain-to-process-emotions

https://my.clevelandclinic.org/health/diseases/21521-psychosomatic-disorder

7

CHAKRAS AND ENERGY CENTERS

Ancient Insight: For thousands of years, Indian and Tibetan traditions have described chakras-energy centers along the spine believed to regulate the flow of life force and influence both physical and emotional health. Blocked chakras were thought to cause both physical and emotional ailments.

Modern Science: While Western science doesn't recognize chakras as physical structures, modern research has found that these ancient energy centers closely align with major nerve plexuses and hormone-producing glands in the body. Practices like acupuncture and meditation, which target these points, have been shown to relieve pain, reduce stress, and boost overall well-being.

The HeartMath Institute's research reveals that our emotions and intentions can create measurable changes in heart rhythm patterns, a phenomenon called "heart coherence." When we practice heart-focused breathing or cultivate positive feelings, our heart, brain, and nervous system become more synchronized, much like the harmonious flow described in ancient energy systems. This state of coherence is linked to reduced stress, better emotional regulation, and improved physical health, echoing the wisdom of holistic traditions through the lens of contemporary science.

Source:

https://www.heartmath.org/research/science-of-the-heart/resilience-stress-and-emotions/

https://www.heartmath.org/research/research-library/basic/heartmath-approach-to-self-regulation-and-psychosocial-well-being/

8

Ayurveda's Circadian Wisdom

Ancient Insight: For thousands of years Ayurveda, the ancient Indian system of medicine, taught that timing is everything - when you eat, sleep, exercise, and work can make or break your health. Ayurveda emphasized the importance of living in harmony with the natural cycles of the day and night.

Modern Science: Ayurveda's ancient wisdom is now being powerfully confirmed by cutting-edge science. Modern chronobiology reveals that our bodies are governed by intricate circadian rhythms that control hormone release, digestion, metabolism, and sleep-wake cycles. Disrupting these rhythms -like eating late at night or irregular sleep - can trigger weight gain, diabetes, and other chronic diseases.

What's truly astonishing is how the body's three fundamental energies follow their own rhythmic cycles throughout the day and seasons, perfectly aligning with these biological clocks. Midday is prime for digestion, early morning is ideal for movement and mental clarity. Ignoring these natural rhythms leads to imbalances, setting the stage for illness.

Today, integrative research shows that adopting Ayurvedic daily and seasonal routines can reset and optimize your internal clock, harmonizing your body, mind, and environment. This ancient science, now backed by Nobel Prize-winning circadian research, offers a revolutionary blueprint for thriving in today's fast-paced world to unlock vitality, prevent disease, and live in sync with the timeless rhythms of life.

Source:
https://lifespa.com/ayurvedic-lifestyle/circadian-rhythm/nobel-prize-validates-ayurvedic-circadian-clock/

9

Gut Health and the Mind

Ancient Insight: Ancient systems like Traditional Chinese Medicine and Ayurveda long emphasized that gut health is the cornerstone of overall wellness, influencing everything from energy levels to emotional balance.

Modern Science: The latest breakthroughs in neuroscience and microbiology uncovered the intricate "gut-brain axis," a dynamic communication network where gut microbes, nerves, and brain constantly interact.

This axis profoundly shapes mood, memory, and mental health by regulating immune responses, hormone levels, and brain plasticity. Recent studies reveal that imbalances in the gut microbiome can contribute to depression, anxiety, and even neurodegenerative diseases, confirming that a healthy gut truly means a healthy mind.

The latest research provides strong support for the ancient insight that gut health is central to overall wellness.

Source:

https://scopeblog.stanford.edu/2025/03/06/gut-brain-connection-long-covid-anxiety-parkinsons/

https://pmc.ncbi.nlm.nih.gov/articles/PMC11853140/

10

FASTING FOR LONGEVITY

Ancient Insight: For thousands of years, cultures worldwide have practiced fasting, often for spiritual and health reasons, believing it could purify both body and mind. From the fasting rituals of ancient Greek philosophers like Pythagoras to religious observances like Ramadan and Yom Kippur, the belief in the benefits of giving the body a break from food has deep roots.

Religious and spiritual leaders like Buddha and even Native American tribes used fasting as a tool for spiritual enlightenment and cleansing the body. They believed that fasting could purify both the body and mind.

Modern Science: Cutting-edge research confirms that intermittent fasting-alternating periods of eating and fasting-can trigger cellular repair processes like autophagy, improve metabolic health, reduce inflammation, and may even extend lifespan.

Recent studies show that fasting can reverse age-related gut dysfunction, support stem cell regeneration, and lower the risk of chronic diseases such as diabetes and heart disease. However, the benefits can vary based on individual genetics and health, and some risks may exist in specific contexts. While the full impact on human longevity is still under investigation, the remarkable alignment between ancient wisdom and modern science highlights fasting as a powerful tool for health and healthy aging.

Source:

https://www.nature.com/articles/s41467-024-45260-9

https://gero.usc.edu/2024/02/20/fasting-mimicking-diet-

https://pmc.ncbi.nlm.nih.gov/articles/PMC8932957/

TRUE! ANCIENT WISDOM, MODERN PROOF

When Old Ideas Turn Out to Be Right After All

Modern science is now confirming what ancient civilizations already seemed to know—whether in medicine, technology, or the natural world. This section explores fascinating cases where time-tested knowledge is being backed up by today's research, revealing just how much the past still shapes the present.

1

Ancient Drills and Natural Fillings: Dentistry Long Before Novocaine

Thousands of years before modern dental clinics, ancient civilizations were already tackling toothaches with surprising skill. As early as 3,000 BCE, the Egyptians and people of the Indus Valley were using stone tools as dental drills to treat cavities and relieve abscesses—long before the invention of anesthesia.

Archaeologists have even found evidence of natural fillings made from mixtures like barley and antibacterial honey, revealing an early grasp of infection control. These treatments weren't just makeshift fixes—they served many of the same purposes as modern materials, impressing today's researchers with how advanced ancient dental care really was.

Source:
https://pubmed.ncbi.nlm.nih.gov/35816771/

2

A Sweet Spot for Natural Healing

The ancient Egyptians didn't just revere honey as a food or to fill holes in their teeth; they used it as a powerful medicine. Archaeologists found pots of honey in tombs that were over 3,000 years old - and still edible. Egyptians believed honey was a gift from the gods, with its magical healing properties. Turns out, they were right.

Modern science has discovered that honey has natural antibacterial, antifungal, and anti-inflammatory qualities. It can even help wounds heal faster. Nowadays, doctors use medical-grade honey to treat burns, cuts, and ulcers, proving that nature's sweet treat is as effective today as it was in ancient times.

Source:
https://www.gentledental-mi.com/ancient-dentistry

https://historyfacts.com/world-history/fact/archaeologists-have-found-3000-year-old-pots-of-honey-that-are-still-edible/

3

Nature's Aspirin: Willow Bark

Imagine being in ancient Greece with a headache. Instead of reaching for a pill, you'd simply head outside, break off some willow bark, and chew it for relief. Hippocrates, the "father of medicine," recommended chewing on willow bark to relieve pain and fevers. This simple natural remedy was used for centuries, especially by Native Americans who brewed it into tea.

Fast forward to the 19th century, and scientists discovered the active ingredient in willow bark—salicylic acid—which later became aspirin, one of the most widely used drugs in the world. Ancient healers didn't know the chemical name, but they certainly knew that it worked.

Next time you pop an aspirin, remember that it's basically ancient wisdom from nature in a pill.

Source:
https://www.sciencehistory.org/stories/magazine/aspirin-turn-of-the-century-miracle-drug/

4

Acupuncture: Ancient Healing Needles

Thousands of years ago in China, healers believed that tiny needles could help balance the body's energy, or "Qi." This led to the development of acupuncture, where needles are carefully inserted at specific points along the body. The idea was that when the body's energy flowed properly, people felt better.

Today, modern science has validated many of these claims—acupuncture can stimulate nerves, muscles, and connective tissue, and it's been shown to release the body's natural painkillers, reduce inflammation, and improve overall health.

Acupuncture isn't just about tiny needles; it's about balancing energy and promoting natural healing. Whether you're using it to relieve pain or reduce stress, acupuncture connects us to an ancient world of healing techniques that continue to thrive in today's high-tech world.

Source:
https://www.nccih.nih.gov/health/acupuncture-effectiveness-and-safety

5

THE MOON AND HUMAN BEHAVIOR

In many ancient cultures, people believed the moon had a powerful effect on human behavior. From wolves howling to people acting strange during a full moon, the moon was thought to influence everything from sleep to sanity. The word "lunatic" actually comes from the Latin word for moon, "luna."

While some dismissed these ideas as superstition, modern research is showing that the moon may indeed impact sleep cycles and behavior. Scientists have found that during a full moon, people tend to sleep less and experience more restless nights—perhaps the moon really does make us a bit wilder.

Next time you're tossing and turning during a full moon, don't blame your mattress—blame the moon. Ancient cultures were onto something when they suspected the moon's mysterious power over humans.

Source:
https://www.sleepfoundation.org/how-sleep-works/do-moon-phases-affect-sleep

https://pmc.ncbi.nlm.nih.gov/articles/PMC7840136/

6

Fermented Foods for Health

Fermented foods like yogurt, kimchi, and sauerkraut have been staples in cultures around the world for thousands of years. Ancient people believed that these foods had special powers to heal the body, especially the gut.

Now, modern science is backing that up. Fermented foods are packed with probiotics—good bacteria that help our digestive system run smoothly. These little microbes also play a huge role in our immune system, metabolism, and even mental health.

Who knew that what's happening in your gut could influence how you feel emotionally? Whether it's ancient sauerkraut or today's kombucha, people have known for centuries that fermented foods make you feel better inside and out.

Source:
https://www.teagasc.ie/about/research--innovation/research-publications/tresearch-articles/2025/more-than-a-gut-feeling.php

7

YOGA: 5,000 YEARS OF MIND-BODY SCIENCE

Yoga isn't just about stretching, it's a 5,000-year-old practice developed in ancient India to bring together body, mind, and breath. Originally designed to support inner balance and mental clarity, yoga has stood the test of time and now modern science is catching up.

Current research confirms that yoga does far more than improve flexibility. It increases levels of calming brain chemicals like GABA and endorphins, which are linked to reduced anxiety and better mood. Studies also show that regular yoga can enhance sleep quality, mental focus, and emotional resilience.

On a physical level, yoga supports muscle strength, balance, and cardiovascular health without placing excessive strain on the body. Practicing yoga in a group setting may also promote social connection, which adds another layer of mental wellbeing.

Best of all, yoga is accessible to nearly everyone regardless of age, fitness level, or body type. Whether you're managing stress, recovering from injury, or just need a mental reset, yoga continues to offer a time-tested path to feeling better, inside and out.

Source:
https://www.hopkinsmedicine.org/health/wellness-and-prevention/9-benefits-of-yoga

8

ANIMAL THERAPY: MAN'S BEST HEALER

For centuries, animals have been considered sources of emotional and spiritual healing. The ancient Egyptians revered cats, believing they had protective and magical properties, while many Native American tribes believed animals were spirit guides.

Today, we see animals playing a crucial role in modern therapy. Research shows that therapy animals, such as dogs, can help to reduce anxiety, stress and depression, and lower blood pressure. Patting a dog has also been proven to release oxytocin, the "feel-good" hormone linked to bonding and emotional well-being.

This modern research confirms what ancient cultures intuited: animals are powerful healers. Regular interaction with therapy animals provides sustained mental health benefits, making them invaluable companions in both clinical therapy and everyday life.

Whether it's a loyal therapy dog or your family pet, animals have an extraordinary ability to heal hearts and calm nerves — just like Egyptians knew thousands of years ago.

Source:
https://www.thezebra.com/resources/research/animal-therapy-statistics/

9

Sound Healing: Music for the Soul

Ancient civilizations like the Greeks and Tibetans used music and sound to heal the body. They believed that different tones and rhythms could balance energy and bring harmony to the body's systems. Tibetan singing bowls, gongs, and chants were used to align energy fields and promote healing.

Fast forward to today, and sound therapy is being used to treat everything from anxiety to chronic pain. The vibrations from sound help sync the brain's rhythm and relax the body, much like ancient sound healers once believed.

Whether it's listening to your favorite song or meditating with a Tibetan bowl, sound has a way of making us feel better. The ancient healers who believed in the power of sound weren't so different from the modern therapists who use it today.

Source:
https://mental.jmir.org/2025/1/e69120

10

LAVENDER FOR CALMING

Long before the invention of aromatherapy, ancient Greeks and Romans used lavender for its calming and soothing properties. They used it in baths and to scent their homes, believing it helped reduce stress and encouraged relaxation. Modern science now confirms that lavender has real, measurable effects on the brain. Studies show that lavender's scent can reduce anxiety, improve sleep, and even ease pain.

Source:
https://jcdr.net/articles/PDF/20432/74832_CE%5BRa1%5D_F(SL)_QC(SD_IS)_PF1(AG_IS)_PFA(SS)_PB(AG_IS)_PN(IS).pdf

FALSE! SCIENCE AND MEDICINE FACTS

Were true and now they are NOT.

Recent research discoveries show how scientific or medical advice previously considered "true" has been overturned or re-evaluated. These facts show how much science can change over time, often overturning widely held beliefs.

1

EGGS - FROM VILLAIN TO HERO

Not so long ago, eggs were the bad guys of the breakfast table. For decades, scientists and health experts warned that eating eggs could raise your cholesterol and increase your risk of heart disease. In the late 1960s, official guidelines even recommended eating no more than three eggs a week, and "low-cholesterol" products filled the supermarket shelves. The fear made sense at the time - eggs are high in cholesterol, and high cholesterol was linked to heart problems.

But here's the twist. Modern research has turned this thinking on its head. We now know that, for most people, the cholesterol you eat in eggs doesn't have much impact on your blood cholesterol or heart health. Recent studies show that even eating a dozen eggs a week doesn't raise cholesterol levels in older adults, and eggs are packed with nutrients like protein, choline, and antioxidants that support your muscles, brain, and eyes.

So, after nearly half a century in the nutritional "naughty corner," eggs are back on the menu for most people. Just remember that what you eat with your eggs such as veggies instead of bacon still matters for your health.

Source:
https://www.monash.edu/medicine/news/latest/2025-articles/regularly-eating-eggs-supports-a-lower-risk-of-cardiovascular-disease-related-death

2

THE FAT MYTH - NOT ALL FATS ARE EVIL

For decades, fat was considered the enemy, and low-fat diets were recommended for everything from weight loss to heart health.

But new research shows it's the type of fat that matters most. Healthy fats-like those in avocados, nuts, seeds, and olive oil-are actually crucial for brain function, hormone production, and may even help protect against diseases like dementia.

Meanwhile, cutting out fat often leads people to eat more sugar and processed carbs, which are now strongly linked to weight gain, heart disease, and other health problems. It turns out your brain loves healthy fats, and the real villain in modern diets is ultra-processed foods loaded with sugar and refined carbs.

Source:
https://www.ucl.ac.uk/news/2025/feb/healthy-fats-could-protect-against-motor-neurone-disease

3

MILK - MAYBE NOT FOR EVERYONE

For decades, milk was hailed as the perfect food for building strong bones. But scientists now know that about two-thirds of the world's population loses the ability to digest lactose-the sugar in milk-after childhood.

This means the majority of adults can't process milk properly, often leading to uncomfortable symptoms like bloating or cramps. Some research even links high dairy intake to issues like acne or digestive troubles.

No wonder plant-based alternatives like oat and almond milk are booming in popularity as they are easier on the gut for many people and offer new options for those who can't tolerate traditional dairy.

Source:
https://www.niddk.nih.gov/health-information/digestive-diseases/lactose-intolerance/definition-facts

4

SALT - FRIEND, FOE, OR JUST MISUNDERSTOOD?

For years, salt has been labeled the bad guy in health advice- cut it down to protect your heart and keep blood pressure in check. And for many people, especially those with high blood pressure, that advice still holds.

But here's the twist: new research shows it's not about banning salt, it's about finding the right balance. Too little salt can be just as risky as too much, especially for people who sweat a lot, like athletes. Salt plays a crucial role in hydration, nerve function, and muscle performance.

The World Health Organization now recommends reducing salt from processed foods and using lower-sodium salt substitutes to improve heart health without cutting salt out entirely.

So, is salt the villain? Not really. For most of us, they key is moderation, context, and lifestyle. Whether you're recovering from a workout or managing your heart health, the smarter move isn't to fear salt, but to understand it.

Source:
https://www.who.int/news-room/fact-sheets/detail/sodium-reduction

5

Vaccines and Autism - The Myth That Won't Die

In 1998, a single study falsely linked vaccines to autism, causing widespread panic and leading many parents to avoid vaccinating their children. This study was later found to be fraudulent and was fully retracted, with the author losing his medical license.

Numerous large-scale studies (e.g. half a million children) have shown no connection between vaccines and autism. Despite this strong body of evidence, the idea of a connection still circulates.

In 2025, the U.S. Centre for Disease Control (CDC) announced it would launch a new study into vaccines and autism in response to ongoing public and political pressure, even though decades of research have already shown no link.

Source:
https://www.cdc.gov/autism/faq/index.html

6

Pluto - Once a Planet, Now a Dwarf

For most of the 20th century, we were taught that Pluto was the ninth planet in our solar system. But in 2006, astronomers redefined the criteria for what makes a planet, and Pluto got demoted to "dwarf planet" status. While it's no longer considered a full-fledged planet, Pluto remains a favorite among space enthusiasts for its icy, mysterious qualities and surprising heart-shaped feature.

Pluto got kicked out of the planetary club but still captures hearts.

Source:
https://science.nasa.gov/dwarf-planets/pluto/

7

Human Brain Cells - Regeneration is Real

It was long believed that once you lost brain cells, they were gone forever. But new research shows that the adult brain can generate new neurons, especially in the hippocampus-the area linked to memory and learning.

A 2024 study found that making new brain cells in adulthood directly supports the ability to remember conversations and learn new information. This discovery is fueling new hope for therapies to restore memory and cognitive function in conditions like Alzheimer's and dementia. Turns out, your brain can grow new cells - so keep those mental workouts going, even as you get older.

Source:
https://stemcell.keck.usc.edu/to-remember-conversations-keep-making-new-brain-cells/

8

Chocolate - Once Banned, Now Beneficial

From sinful treat to heart-helper, dark chocolate's reputation has had a sweet turnaround.

Chocolate was once seen as an indulgence to be enjoyed sparingly, but new research shows that dark chocolate can actually offer real health benefits when you choose the right kind.

Recent large studies have found that people who eat dark chocolate regularly have a lower risk of developing type 2 diabetes and may see improvements in heart health, thanks to its high levels of antioxidants and beneficial plant compounds.

The catch? The benefits are linked to dark chocolate with high cocoa content and low sugar - not milk chocolate or sugary chocolate bars.

Source:
https://hsph.harvard.edu/news/eating-dark-chocolate-linked-with-reduced-risk-of-type-2-diabetes/

9

Eating Late - Not as Bad as We Thought

For years, people were told not to eat after 6 PM to avoid weight gain. But new research from Johns Hopkins suggests that the timing of your meals is less important than what and how much you eat.

Your metabolism doesn't "shut off" at night, and eating a small, healthy snack before bed can be perfectly fine for most people, especially if it helps you sleep or prevents late-night overeating. The key is to focus on healthy food choices and total calories, rather than the clock.

Source:
https://hub.jhu.edu/2024/04/22/study-challenges-intermittent-fasting/

10

Water Consumption - Do You Really Need 8 Glasses?

The classic "8 glasses of water a day" rule has been debunked by modern science. According to the National Academies of Sciences, Engineering, and Medicine, how much water you need depends on your age, activity level, climate, and health.

Most people get enough by drinking when thirsty and paying attention to their body's signals. Plus, you get a significant amount of water from foods like fruits and vegetables-so you don't need to count glasses.

The bottom line: listen to your body, and remember you've been hydrating with watermelon and cucumbers all along.

Source:
https://www.nationalacademies.org/our-work/dietary-reference-intakes-for-electrolytes-and-water

11

Coffee - From Villain to Hero

Coffee was the bad guy for its supposed link to heart disease. Now coffee is celebrated for its surprising health benefits.

Recent studies show that people who drink coffee regularly, especially in moderate amounts of 2 to 3 cups per day, tend to have a lower risk of heart disease, Parkinson's disease, Alzheimer's, type 2 diabetes, and even some types of cancer.

Coffee's antioxidants and anti-inflammatory compounds are thought to play a key role, and coffee drinkers often live longer than those who skip their daily cup. Just remember: the benefits are best when you keep added sugars and heavy creamers to a minimum, and those with very high blood pressure should still be cautious.

Source:
https://www.heart.org/en/news/2022/08/08/is-caffeine-a-friend-or-foe

12

"Healthy" Margarine - Was Worse Than Butter

In the 1980s and 90s, people were encouraged to swap butter for margarine, believing it was better for heart health. Early margarines did contain trans fats, which we now know increase the risk of heart disease.

However, most margarines today have been reformulated and no longer contain trans fats. Modern soft or tub margarines are made with healthier unsaturated oils and are generally a better choice for heart health than butter, which is high in saturated fat and can raise LDL cholesterol.

For the healthiest option, choose soft or liquid margarines with minimal saturated fat and avoid stick margarines and spreads with added trans fat. Butter can still be enjoyed in moderation, but it's no longer seen as the healthier option.

Source:
https://newsroom.heart.org/news/replacing-butter-with-plant-based-oils-may-reduce-the-risk-of-premature-death

13

Vitamin C and Colds: Myth Busted

For years, many believed that taking large doses of Vitamin C could prevent or cure the common cold.

Recent studies show that while regular Vitamin C supplementation may slightly reduce how long a cold lasts and make symptoms milder, it does not prevent most people from catching a cold. The exception is for those under heavy physical stress, where Vitamin C can cut the risk in half.

While your orange juice is still great for overall health, it's not the cold cure-all we once thought.

Source:
https://pubmed.ncbi.nlm.nih.gov/39803741/

14

Stretching Before Exercise? Maybe Not

Holding your stretch before a big race? For years, athletes were told to stretch before exercising to prevent injury.

But recent research shows that static stretching (holding a position) before a workout can actually decrease strength and performance, especially if stretches are held for longer than 30 seconds. Instead, dynamic stretching where you move while stretching is now recommended to warm up muscles, improve blood flow, and get your body ready for action.

Science now says static stretching still has a place, but it's best saved for after exercise to help with flexibility and recovery. Skip the long holds before you move - save them for your cooldown.

Source:
https://health.clevelandclinic.org/dynamic-stretching-vs-static-stretching

15

Sugar Makes Kids Hyper? Not Really.

It's a belief that has been passed down for generations -sugar makes kids hyper. But rigorous research conducted by experts has consistently failed to find a connection between sugar and hyperactivity. A 2024 study summarizes decades of placebo-controlled studies and meta-analyses, all showing that sugar consumption does not lead to hyperactivity or disruptive behavior in children, even in those diagnosed with ADHD.

Instead, it's often the excitement of parties or events where sugary foods are served that makes kids act out. While sugar may provide a temporary energy boost, it does not turn children into hyperactive whirlwinds.

Parental expectations can also play a role in how children's behavior is perceived.

While sugar isn't the culprit for hyperactivity, it is still linked to other health concerns, such as obesity and dental issues, so moderation is important.

Source:
https://thesector.com.au/2024/06/01/no-sugar-doesnt-make-your-kids-hyperactive/

16

Carrots Don't Help You See in the Dark

We've all heard that eating carrots will help you see in the dark. While it's true that carrots are good for eye health because they provide vitamin A, they won't give you super sight.

Unless you are deficient in vitamin A, eating more carrots won't improve your vision or let you see in the dark.

The myth that carrots boost night vision began during World War II, when British pilots' success at night was attributed to eating carrots - a story spread by the British Government to hide the use of secret radar technology.

Carrots are healthy, but they won't turn you into a fighter pilot with night-vision goggles.

Source:
https://www.smithsonianmag.com/history/carrots-cant-help-you-see-in-the-dark-heres-how-world-war-ii-propaganda-campaign-popularized-the-myth-28812484/

17

No - Cracking Your Knuckles Won't Cause Arthritis

For years, parents warned kids that cracking their knuckles would lead to arthritis. But the science says that is just a myth.

The popping sound you hear when you crack your knuckles actually comes from tiny gas bubbles bursting in the joint fluid - not from bones grinding or damage being done. Interestingly, once you crack a knuckle, you can't make the same popping sound again for about 20 minutes. That's because it takes time for the gas bubbles to build up again in the joint fluid, ready for the next satisfying pop.

While knuckle cracking might annoy those around you, it's harmless and won't cause arthritis.

Source:
https://www.webmd.com/osteoarthritis/joint-cracking-osteoarthritis

18

Fat Doesn't Always Make You Fat

The belief that "eating fat makes you fat" has been debunked. Now we know that healthy fats, like those found in nuts, avocados, and olive oil, are crucial for brain function and weight management. It's refined carbohydrates and sugar that are more likely to cause weight gain.

Don't fear the fat - your brain loves it. It's sugar and carbs that are the sneaky villains in weight gain.

Source:
https://newsinhealth.nih.gov/2019/07/skinny-fat

19

Shaving Doesn't Make Hair Grow Back Thicker

Shaving doesn't turn you into a werewolf - it just feels that way because of those blunt ends. Many people believe that shaving hair makes it grow back thicker and darker. But this myth has been debunked: shaving doesn't affect hair growth, thickness, or color. The stubbly feeling is just because the hair is cut bluntly at the skin surface.

Source:
https://www.mayoclinic.org/healthy-lifestyle/adult-health/expert-answers/hair-removal/faq-20058427

20

THE CLASSIC FOOD PYRAMID – WRONG

Remember the classic food pyramid, with bread, rice, and pasta forming the base? Nutrition experts now agree that model was outdated as it encouraged too many refined carbs and not enough healthy fats or proteins.

The latest guidance from Harvard's Healthy Eating Pyramid and Healthy Eating Plate (2025) recommends filling your plate with vegetables, fruits, healthy fats (like olive oil, avocado and nuts), and lean proteins (such as beans, fish, and chicken). Refined grains and processed foods should be limited, while whole grains are preferred.

Source:
https://nutritionsource.hsph.harvard.edu/healthy-eating-plate/

21

THE 5-SECOND RULE IS A MYTH - BUSTED

Sorry, but food dropped on the floor isn't magically safe to eat if you pick it up within five seconds. Bacteria can contaminate food almost instantly.

Rutgers University research shows that the risk of contamination depends on the type of food (moist foods pick up more bacteria), the surface it lands on (carpet transfers less than tile or steel), and how long it sits there (longer time means more bacteria). So, even though it's tempting, think twice before applying the "5-second rule" to that fallen cookie.

Source:
https://www.rutgers.edu/news/rutgers-researchers-debunk-five-second-rule-eating-food-floor-isnt-safe

22

Humans – the sensory Superstars

Most of us were taught that humans have five senses. However, scientists now know we have at least nine - and some experts count up to 21 or more. Beyond the classic five of taste, touch, smell, sight and sound, you also sense balance (so you don't topple over), temperature (hot and cold), pain, and even internal signals like hunger and body awareness (proprioception). There's even a "day/night" sense that helps regulate your sleep cycle. So, your body is constantly gathering way more information than you might realize.

Source:
https://www.sciencefocus.com/the-human-body/how-many-senses-do-we-have

History's Hidden Surprises: Quirky Finds That Rewrite the Past

Forget everything you learned in school. History is full of unexpected twists, oddball inventions, and strange moments.

Here you will meet emus that outsmarted the military, ancient "your mom" jokes, vampire-proof graves, and even a flood of beer through the streets of London. Each quirky fact is backed by recent discoveries, offering a fresh look at the past that will leave you amused, amazed, and eager to dig deeper into the world's weirdest true stories.

1

THE GREAT EMU WAR OF 1932

In Australia, humans once went into a bizarre battle with emus— the giant flightless birds. In 1932, farmers were overwhelmed by emu invasions that destroyed crops, so the military was called in to fight them off. The emus were remarkably fast, able to outrun the soldiers and machine guns, and their zigzagging running patterns made them difficult to hit. The military operation lasted for several weeks but failed to significantly reduce the emu population and the military withdrew.

Source:
https://www.britannica.com/topic/Emu-War

2

The Year Without a Summer

In 1816, the world experienced an unusual phenomenon known as the "Year Without a Summer." A massive volcanic eruption in Indonesia spewed ash into the atmosphere, leading to global cooling. Crops failed, and snow fell in the summer months. People were forced to invent new recipes, leading to some truly unusual dishes—like "snow soup."

Source:
Smithsonian

https://www.smithsonianmag.com/history/blast-from-the-past-65102374/

3

Napoleonic Rats

During Napoleon's campaign in Egypt and Syria, his soldiers faced more than just enemy fire—they also battled disease-spreading rats and fleas. Poor hygiene and unsanitary camp conditions created a perfect storm for outbreaks of illnesses like the bubonic plague, which claimed many lives. It turns out that in the chaos of war, tiny enemies can be just as dangerous as armed ones.

Source:
https://www.montana.edu/historybug/napoleon/plague-syria.html

4

The First "Scavenger Hunt"

In 1927, an unusual contest was held in Paris that can be seen as the world's first scavenger hunt. Participants were challenged to collect a series of bizarre items, including a piece of hair from a Frenchman and a bottle of champagne. The contest was a clever marketing ploy to promote a new line of women's clothing, but it also inspired countless scavenger hunts that followed.

Source:
https://ruinedtable.substack.com/p/elsa-maxwell

5

The Vampire Graveyard of Poland

In 2022, archaeologists in Poland uncovered a 17th-century graveyard near the village of Pien, revealing over 100 unmarked burials exhibiting "deviant" burial practices intended to prevent the dead from rising. Among the most striking discoveries was the skeleton of a young woman, dubbed "Zosia," buried with a sickle placed over her neck and a padlock on her toe—measures believed to stop her from returning as a vampire. These findings align with historical accounts of Eastern European vampire lore, where such practices were used to protect the living from the undead.

Source:
https://nypost.com/2024/10/28/science/real-life-vampire-discovered-padlocked-and-pinned-inside-grave/

https://www.atlasobscura.com/articles/anti-vampire-graves-poland

Young woman 'vampire' buried with a sickle across her neck to cut off her head if she tried to rise again.

6

The First Computer Bug

In 1947, engineers working on the Mark II computer at Harvard University found a moth stuck in one of the components. They taped the insect into their logbook and labeled it "first actual case of bug being found." The words "bug" and "debug" soon became a standard part of the language of computer programmers.

Source:
https://americanhistory.si.edu/collections/object/nmah_334663

7

The Oldest Known "Your Mom" Joke

Did you know the oldest recorded "your mom" joke dates back to 1500 B.C. in ancient Babylon? It was inscribed on a tablet, and while the exact wording is lost to time, it proves that humor and playful insults have been around for a long time. Other jokes and riddles in the tablet's inscriptions touch on topics including sex, politics, and beer. Apparently, ancient Babylonians knew how to have a laugh too.

Source:
https://allthatsinteresting.com/historys-first-yo-mama-joke

8

THE CAT PIANO

In the 18th century, a quirky inventor proposed a strange musical instrument called the "cat piano." It was said to be a keyboard where each key was attached to a cat, and when pressed, the cats would meow. Although it was never actually made, the idea tickled the fancy of many and proves that some music trends should never be followed.

Source:
https://www.smithsonianmag.com/smart-news/music-or-animal-abuse-brief-history-cat-piano-180963056/

9

THE DANCING PLAGUE OF 1518

In July 1518, residents of Strasbourg, France, experienced a mysterious phenomenon known as the Dancing Plague. Hundreds of people danced uncontrollably for days on end, leading to exhaustion and even death. Historians believe it might have been a mass hysteria caused by stress or ergot poisoning (a hallucinogenic mold).

Source:
https://www.history.com/articles/what-was-the-dancing-plague-of-1518

10

THE LONDON BEER FLOOD

In 1814, a huge vat of beer at the Meux and Company Brewery in London ruptured. The force of the liquid destroyed the brewery's rear wall and flooded the surrounding area of St Giles, demolishing two houses and damaging others, causing a flood of over 323,000 liters of beer to spill into the streets. The incident caused considerable chaos. While some residents were thrilled to see the streets running with beer, others were less enthusiastic about cleaning up the aftermath.

Source:
https://www.historic-uk.com/HistoryUK/HistoryofBritain/The-London-Beer-Flood-of-1814/

Quiz: History's Hidden Surprises

Fill in the Blank

Fill in the blanks below to test your quirky history knowledge!

1. In 1932, the Australian military went to war with __________ — and lost!

2. The "Year Without a Summer" was caused by a massive __________ eruption in Indonesia.

3. Napoleon's troops had to battle not just enemies, but also giant __________ that stole their food and invaded their camps.

4. In 1927, Paris hosted the world's first modern-day __________ hunt, where participants collected bizarre items like hair and champagne.

5. A 5,000-year-old sandwich found in an Egyptian tomb contained stale bread and mouldy __________.

6. The first "computer bug" was actually a __________ found in a computer in 1947.

7. The oldest recorded "your mom" joke dates back to ancient __________ in 1500 B.C.

8. The "cat piano" was a weird 18th-century invention where __________ were supposed to meow when keys were pressed.

9. The Dancing Plague of 1518 saw people in Strasbourg dance uncontrollably, possibly due to __________ poisoning.

10. In 1814, London streets were flooded with over 300,000 litters of __________ after a brewery vat burst.

Answers

1. Emus
2. Volcanic
3. Rats
4. Scavenger
5. Cheese
6. Moth
7. Babylon
8. Cats
9. Ergot
10. Beer

Brilliant & Bold: Women Shaping Science Today

These real-life stories celebrate female scientists pushing boundaries and breaking ground in their fields. From decoding the brain to fighting disease, these women are turning curiosity into impact and lighting the way for the next generation of changemakers.

1

THE ICE WHISPERER: JULIE BRIGHAM-GRETTE

Imagine being a superhero for our planet. Dr. Julie Brigham-Grette is doing just that as she studies the Antarctic ice sheets. Her groundbreaking work reveals that these massive icebergs are melting faster than we thought. Julie uses ancient ice records to tell us how climate change can affect our world. By uncovering the mysteries of ice, she's helping us understand what we need to do to protect our Earth. If you've ever dreamed of saving the planet, take a cue from Julie. You too can make a difference.

Source:
https://en.wikipedia.org/wiki/Julie_Brigham-Grette

2

THE MATH WHIZ FIGHTING INFECTIOUS DISEASE: ALISON GALVANI

Math and science can help save millions of lives. Dr. Alison Galvani uses powerful computer models to understand how infectious diseases like COVID-19, Ebola, and influenza spread through populations. By simulating outbreaks, she helps governments and health organizations make smarter decisions about vaccines and public health policies. Her work shows how science can protect communities and stop epidemics before they start. Inspired by a passion for helping others, Alison proves that behind every number is a chance to make the world healthier and safer.

Source:
https://ysph.yale.edu/profile/alison-galvani/

3

The Quantum Challenger Rewriting the Rules of Computing: Ewin Tang

At just 18, Ewin Tang shook up the world of quantum computing by proving that classical computers can solve some problems just as fast as quantum ones—something experts thought was impossible! Her breakthrough "dequantizing" quantum algorithms means regular computers might be more powerful than we imagined, changing how scientists approach machine learning and quantum research. Ewin's work shows that sometimes, challenging big ideas can lead to even bigger discoveries—and that curiosity and creativity are the real superpowers in science.

Source:
https://www.quantamagazine.org/teenager-finds-classical-alternative-to-quantum-recommendation-algorithm-20180731/

4

The Bee Whisperer: Christina Grozinger

Buzz, buzz. Did you know these little creatures can recognize human faces? Christina's research dives into the complex social lives of bees, showing us just how smart they really are.

She also investigates how stressors like climate change, pesticides, and poor nutrition are driving global pollinator declines, and her work has shown that bees are more resilient when they have access to diverse, high-quality food sources1. By combining genomics, physiology, and ecology, Christina Grozinger develops practical strategies to support bee health and guides efforts to restore pollinator habitats in both farms and wild landscapes. By studying their behavior, she's advocating for their protection, which is essential for our planet.

Source:
https://pollinators.psu.edu/about/directory/christina-grozinger-ph-d

5

The Celiac Code Cracker: Cisca Wijmenga

Dr. Cisca Wijmenga is a global expert in the genetics of celiac disease. She identified 39 genetic risk factors that help explain why some people develop this autoimmune condition. Her research has also shown that celiac disease shares genetic links with other immune disorders like type 1 diabetes, rheumatoid arthritis, and Crohn's disease. By uncovering these connections, Dr. Wijmenga is helping scientists move toward earlier diagnoses and more effective treatments for millions of people worldwide.

Source:
https://www.rug.nl/research/genetics/staff/cisca-wijmenga?lang=en

6

The Deep Diver Heating Up Gene Editing: Jennifer Doudna

Imagine having tiny genetic scissors that can precisely cut and edit DNA to fix diseases—that's exactly what Nobel prize winner Dr. Jennifer Doudna helped create with her groundbreaking discovery of CRISPR-Cas9. Now, Jennifer is exploring some of the hottest places on Earth. Her team recently studied GeoCas9, a remarkable enzyme discovered in bacteria that thrive in scorching hot springs and deep-sea hydrothermal vents. They are looking for even better tools that could make gene editing safer, faster, and more accessible for everyone. Her work is opening doors to treatments for genetic disorders, cancer, and beyond. It shows how curiosity and innovation can turn the secrets of nature into breakthroughs that could transform medicine for everyone.

Source:
https://nihrecord.nih.gov/2024/06/21/doudna-describes-exciting-future-crispr

Source: https://gladstone.org/news/crispr-powered-cancer-shredding-technique-opens-new-possibility-treating-most-common-and

7

The HIV Heroine: Anna Wald

Dr. Anna Wald is a global leader in the fight against genital herpes—a virus that quietly affects over 500 million people worldwide and also plays a hidden but powerful role in the HIV epidemic. Her groundbreaking research revealed why people with genital herpes are up to three times more likely to acquire HIV. This discovery has major implications for public health, especially in regions where both viruses are common, and highlights how tackling herpes could be a key strategy in reducing HIV's global impact. Dr. Wald's work shows that solving one health puzzle can unlock answers to even bigger challenges.

Source:
https://www.fredhutch.org/en/faculty-lab-directory/wald-anna.html

8

The Trailblazing Fossil Hunter Who Unearthed Earth's Ancient Secrets: Mary Anning

Mary Anning was the pioneering fossil hunter from England's Jurassic Coast, whose discoveries changed the world of paleontology forever. As a young woman in the early 1800s, When Mary was just 12 years of age, she unearthed the first complete Ichthyosaurus skeleton and later discovered the first intact Plesiosaurus and Britain's first Pterosaur. She even helped scientists understand "coprolites"—fossilized dino poop! Despite facing discrimination and being excluded from scientific societies because she was a woman, Mary's sharp eye and relentless curiosity revealed ancient sea monsters and inspired generations of scientists. Today, her fossils are displayed in major museums, and her legacy proves that passion and perseverance can rewrite history—even if you start out searching for seashells on the beach.

Source:
https://www.nhm.ac.uk/discover/mary-anning-unsung-hero.html

9

The Whale and Drone Connector: Vanessa Pirotta

Vanessa is famous for using drones to collect "whale snot"—the spray from a whale's blowhole—to study whale health. This innovative, non-invasive technique has taken her from Antarctica to Tonga and Madagascar, helping scientists learn more about these ocean giants without disturbing them. Beyond whales, Vanessa also leads projects to detect illegal wildlife trafficking using AI and works with Indigenous rangers to protect marine life in Sydney Harbour. She's not just a scientist—she's a passionate science communicator, children's book author, and a true champion for wildlife conservation.

Source:
https://www.vanessapirotta.com/

10

The Gut-Brain Revolution: Emma Allen-Vercoe

Dr. Emma Allen-Vercoe is a trailblazing scientist who's made it her mission to uncover the hidden world of our gut microbiome—and how it shapes our health in surprising ways. She invented the "Robogut," a mechanical colon that recreates the unique environment of the human gut, allowing her team to grow and study the bacteria that most scientists once thought were impossible to culture.

Emma's research has revealed that these tiny microbes don't just help us digest food; they may also play a crucial role in mental health, immunity, and even conditions like autism and cancer. By showing that gut bacteria can affect our mood and well-being, she's opened the door to new therapies—like using probiotics tailored to your personal microbiome. Emma wants everyone to see themselves as "guardians" of their own microbes.

Source:
https://researchcbs.ca/faculty/emma-allen-vercoe/

Modern Minds: Meet the Men Behind Many of Today's Science Breakthroughs

These real-life stories spotlight male scientists making waves in the world of discovery from coding late at night to studying emotions in machines and so much more. You'll get a glimpse of the latest innovations and the curious, determined humans driving them forward.

1

THE ROBOT WHISPERER: ALAN WINFIELD

Alan Winfield doesn't just build robots—he teaches them kindness. In his latest project, Winfield worked on creating robots that can "feel" empathy. For example, one robot learned to prioritize helping others, even at its own expense. Imagine a future where robots save people in danger. Alan believes that robots should follow a moral code, making the world safer for everyone. His lab is filled with creative chaos, from robot prototypes to colorful sketches of their "personalities." Winfield's work is paving the way for robots as not just machines but ethical partners in our world.

Source:
https://people.uwe.ac.uk/Person/AlanWinfield

2

DEEP-SEA DAREDEVIL: VICTOR VESCOVO

Victor Vescovo has mapped parts of the ocean no one else has ever seen. Recently, he piloted a submersible into the Mariana Trench, Earth's deepest underwater chasm. During one mission, his team discovered species of translucent jellyfish and worms adapted to crushing pressure. What's more, Vescovo uncovered human trash at these depths, sparking global conversations about pollution. When he's not in the ocean, he's climbing the world's highest peaks, proving that exploration knows no boundaries—above or below.

Source:
https://en.wikipedia.org/wiki/Victor_Vescovo

3

The Quantum Kid: Alexei Kitaev

Alexei Kitaev's work on quantum computers might sound like science fiction, but it's real—and groundbreaking. His research focuses on how particles called "anyons" can store and process vast amounts of information faster than traditional computers. Quantum computers may one day solve unsolvable problems, from curing diseases to understanding climate change. Kitaev is a quiet genius who loves puzzles and views quantum computing as the ultimate brain teaser.

Source:
https://heritageproject.caltech.edu/interviews-updates/alexei-kitaev

4

Gene Detective: Feng Zhang

Feng Zhang is a pioneering scientist who helped develop CRISPR-Cas9. This revolutionary gene-editing tool allows researchers to precisely modify DNA in living cells. His lab continues to advance gene-editing technology, making it more accurate and expanding its medical applications—including the development of the first FDA-approved CRISPR-based therapy for sickle cell disease in 2023. Zhang's innovations are accelerating biomedical research and bringing new hope for treating diseases like sickle cell, beta thalassemia, and even some neurodevelopmental disorders. Inspired by curiosity from a young age, Zhang's work is not only transforming medicine but also sparking important conversations about the future of genetic technology.

Source:
https://en.wikipedia.org/wiki/Feng_Zhang

5

THE BIRD-MAN WHO BEAT TIME: TIM BIRKHEAD

Tim Birkhead studies birds, but he's also cracked mysteries of their ancient past. His recent discovery? Birds evolved the ability to lay eggs with different shell colors millions of years ago to protect them from predators.

He is a renowned British ornithologist whose latest research explores the deep history of our relationship with birds, as well as the mysteries of bird reproduction and egg colors. His new book Birds and Us: A 12,000 Year History, from Cave Art to Conservation, investigates how humans' attitudes toward birds have evolved from seeing them as resources to recognizing their need for protection. He is also famous for revealing that egg color and pattern may signal the health of the mother as well as protect eggs from predators.

Source:
https://cannonballread.com/2020/06/the-most-perfect-thing-inside-and-outside-a-birds-egg-kimmie/

6

The Time Traveler: Kip Thorne

Have you ever dreamed of bending time and space? Kip Thorne helped prove that gravitational waves—ripples in the fabric of space-time—exist. This discovery confirmed one of Einstein's theories and won him the Nobel Prize. Thorne's work was so visionary it inspired the science behind the movie Interstellar. Imagine working on math equations that could one day help us explore distant galaxies.

Source:
https://en.wikipedia.org/wiki/Kip_Thorne

7

Mushroom Mastermind: Paul Stamets

Paul Stamets believes mushrooms can save the world, and he's proving it. His recent discovery of fungi that can eat plastic is a game-changer for the environment. Stamets' lab looks like a sci-fi movie set, filled with glowing mushrooms and experiments on sustainable farming. As a boy, he once got lost in the woods and found his way back thanks to a glowing mushroom—a magical moment that started his lifelong obsession.

Source:
https://www.discovermagazine.com/environment/how-mushrooms-can-save-the-world

8

The Computer Scientist Who Gave AI Imagination: Ian Goodfellow

Ian Goodfellow's invention of GANs (Generative Adversarial Networks) happened in a flash of inspiration one night, when he coded the idea after a casual conversation with friends - and it worked on the first try. He is a brilliant AI pioneer whose invention of GANs was a breakthrough that gave machines a kind of imagination by having two neural networks compete and learn from each other. He started as a Stanford undergrad working with AI legend Andrew Ng, then went on to earn a PhD under Yoshua Bengio, one of the "godfathers" of AI. His work now powers advances in image generation, machine learning security, and AI ethics, showing how curiosity and persistence can lead to game-changing innovation that shapes the future.

Source:
https://www.technologyreview.com/2018/02/21/145289/the-ganfather-the-man-whos-given-machines-the-gift-of-imagination/

9

Plastic Buster: Boyan Slat

At just 18, Boyan Slat created The Ocean Cleanup, an invention designed to remove plastic waste from the oceans. His most recent achievement? Collecting over 200,000 pounds of trash from the Pacific Garbage Patch. Boyan was inspired as a teenager after seeing plastic pollution during a scuba diving trip. Now, he's a hero for marine animals—and for the planet.

Source:
The Ocean Cleanup

10

The Solar Superhero: Saul Griffith

Saul Griffith is on a mission to make clean energy accessible to everyone. His latest project involves creating solar panels that can power entire neighborhoods at a fraction of the cost. Saul is a big believer in practical solutions, like using everyday tools to make cutting-edge tech. As a boy, he was always taking apart gadgets to see how they worked. Now, he's using that curiosity to save the Earth.

Source:
https://www.saulgriffith.com/

Prehistoric Showstopper Fun Facts: Strange Creatures and Surprising Secrets from the Lost World

Journey back in time to encounter some of the most astonishing creatures ever to roam the planet, many only recently discovered or understood by paleontologists.

Their recent breakthroughs are rewriting what we know about Earth's ancient creatures. Here we explore some of the wildest and most unexpected fossil finds, offering a fresh look at the bizarre adaptations and surprising lifestyles that shaped life long before humans appeared.

1

Prehistoric Sea Monsters with Spiral Jaws

The Helicoprion was a terrifying prehistoric shark-like creature that lived 270 million years ago, before dinosaurs. This sea creature had a bizarre spiral jaw that looked like a circular saw. This "buzzsaw shark" could slice through its prey with rows of spiraling teeth. It took over a century for palaeontologists to figure out what that jaw was for. The teeth fossil was found in Western Australia.

Source:
https://visit.museum.wa.gov.au/learn/news-stories/fossil-so-weird-it-stumped-scientists-decades-huge

Credit: Scientific American

2

POOP AND VOMIT FOSSILS REVEAL MORE THAN BONES AND TEETH DO

Innovative use of high-resolution imaging technology is turning dollops of dung fossils into goldmines of information. The remains of bygone meals provide an intimate peek at how small dinosaurs eventually displaced rivals, diversified into myriad shapes and sizes, and grew into giants between 230 million and 200 million years ago in the Polish basin region.

Source:
https://www.science.org/content/article/dino-droppings-reveal-how-giant-beasts-came-dominate-earth

3

GIANT ARMADILLOS ONCE ROAMED SOUTH AMERICA

Imagine an armadillo the size of a small car. The Glyptodon was an ancient mammal that roamed South America during the Ice Age. These massive creatures weighed up to 2,000 pounds and had a shell that could rival a tank's armor. Researchers have evidence of butchery 20,000 years ago on their fossils from Argentina.

Source:
https://www.sciencedaily.com/releases/2024/07/240717162440.htm

4

The Saber-Tooth Cat Had a Weak Bite—But Deadly Teeth

Though they were fearsome predators, saber-tooth cats actually had a surprisingly weak bite compared to modern lions. Instead, their long, curved teeth were designed for precision strikes, targeting soft parts of their prey, like the throat. It's like having a knife instead of a hammer for hunting.

Source:
https://www.nationalgeographic.com/science/article/sabre-toothed-cats-had-weak-bites

5

Woolly Mammoths Had Earmuffs (Sort of)

Woolly mammoths weren't just covered in fur—they also had tiny ears compared to modern elephants. Their small ears helped prevent heat loss in the icy conditions of the Ice Age, almost like built-in earmuffs. So, the next time you bundle up, think of the mammoths who rocked their own winter accessories.

Source:
https://www.popsci.com/environment/woolly-mammoths-evolution/

6

Prehistoric Crocodiles Were Fast Land Runners

Not all prehistoric crocodiles lounged in swamps—some, like Kaprosuchus, were terrifying land predators that could run on their hind legs. Imagine a 20-foot-long croc sprinting after you on dry land. They even had boar-like tusks, earning them the nickname "boar croc".

Source:
https://animals.howstuffworks.com/dinosaurs/kaprosuchus.htm

7

Ancient Penguins Were Giants

Modern penguins are cute, but their prehistoric relatives were enormous. Kumimanu, an ancient penguin, stood over 5 feet tall and weighed around 200 pounds—about the size of a grown human. These giant penguins roamed the seas around New Zealand 60 million years ago, but sadly, no tuxedo-wearing birds of this size are alive today.

Source:
https://www.smithsonianmag.com/smart-news/fossils-of-a-340-pound-giant-penguin-found-in-new-zealand-180981611/

8

GIANT SLOTHS ONCE DUG MASSIVE UNDERGROUND TUNNELS

Around 10,000 years ago, giant ground sloths the size of elephants dug enormous tunnels in South America that can still be seen today. These tunnels, known as "paleoburrows," stretch for hundreds of feet and are large enough to walk through. It's like having an underground subway system—designed by sloths.

Source:
https://www.bbc.com/travel/article/20231127-brazils-mysterious-tunnels-made-by-giant-sloths

9

FISH WITH LEGS?

The Tiktaalik, a fish that lived around 375 million years ago, had limb-like fins that allowed it to crawl out of the water onto land. This incredible creature was one of the first to make the transition from water to land, paving the way for all land animals today—including humans.

Source:
https://www.nature.com/news/2006/060403/full/news060403-7.html

10

50-Million-Year-Old Ants Were Way Bigger

Ants might be small today, but 50 million years ago, there were giant ants the size of hummingbirds. Fossils of these ancient ants were discovered in Wyoming, showing that the insects could grow to be over 2 inches long. Imagine spotting one of those in your picnic basket.

Source:
https://www.bbc.com/news/science-environment-13269302

Far Out Fun Facts from Outer Space

Strap in to discover the latest and greatest mind-blowing wonder facts of the cosmos that prove truth is often stranger - and more spectacular - than science fiction.

In recent years, astronomers and space probes have uncovered astonishing new details about our solar system and beyond. Thanks to cutting-edge missions like the James Webb Space Telescope and NASA's Mars rovers, we're learning more than ever about the wild, mysterious, and sometimes downright bizarre nature of the universe.

1

The Sun - Our Mighty Neighborhood Star

The Sun is much more than just a bright spot in our sky—it's a colossal star at the heart of our solar system. If you could hollow it out, you'd be able to fit about 1.3 million Earths inside.

This giant ball of glowing gas, made mostly of hydrogen and helium, is the powerhouse that makes life on our planet possible. The Sun's energy drives Earth's climate, warming our world, powering weather patterns, and fueling the water cycle. While the Sun's brightness and activity naturally fluctuate, bringing sunspots and dramatic solar flares that can disrupt satellites and communications, these changes have only a tiny effect on Earth's long-term climate.

The recent rapid global warming we see today is driven by human activities, not the Sun. Still, the Sun's steady glow remains essential, shaping the rhythms of our planet and lighting up our days.

Source:
https://science.nasa.gov/sun/

2

Venus – A Day Lasts Longer Than a Year

Venus is a planet of extremes, and one of its strangest features is its day. On Venus, a single day (the time it takes to spin once on its axis) is actually longer than its entire year (the time it takes to orbit the Sun).

A Venusian day lasts about 243 Earth days, while a year on Venus is just 225 Earth days. This oddity is due to Venus's incredibly slow and backwards rotation. If you could stand on the surface (and somehow survive the crushing heat and toxic atmosphere), you'd see the Sun rise very slowly in the west and set in the east. It's a cosmic reminder that not all planets play by the same rules as Earth.

Source:
https://en.wikipedia.org/wiki/Venus

3

Olympus Mons - The Towering Giant on Mars

Mars is home to the largest volcano in the solar system. Olympus Mons is a mountain so massive it defies earthly comparison.

Rising about 13.6 miles (22 kilometers) high, Olympus Mons stands nearly two and a half times taller than Mount Everest, Earth's highest peak. Its base stretches an astonishing 373 miles (600 kilometers) across, covering an area about the size of Italy or Arizona. This colossal shield volcano formed over millions of years as lava slowly piled up, undisturbed by shifting tectonic plates like those on Earth. If you stood on its gentle slopes, you might not even realize you were climbing a mountain, but from space, Olympus Mons is a breathtaking Martian landmark—an enduring testament to the Red Planet's wild volcanic past.

Source:
https://science.nasa.gov/mars/facts/

4

Saturn - The Giant That Could Float

Saturn, famous for its stunning rings, is truly unique among the planets of our solar system. This is not just for its beauty, but for its surprising lightness. Despite being more than 95 times as massive as Earth and nine times wider, Saturn is made mostly of hydrogen and helium, making it less dense than water. In fact, if you could find a bathtub big enough, Saturn would actually float! Of course, this is only a fun thought experiment. There's no ocean in the universe large enough to hold it, and Saturn's gravity would quickly ruin any cosmic bath. Still, Saturn's ability to float is a testament to its status as the solar system's lightest giant, and a reminder that in space, things aren't always what they seem.

Source:
https://science.nasa.gov/saturn/facts/

5

THE MOON IS SLOWLY DRIFTING AWAY

Every year, the Moon edges a little farther from Earth at a pace similar to the growth of your fingernails.

This slow-motion separation is caused by the tug of Earth's tides, which transfer energy to the Moon and push it outward in its orbit. Over millions of years, this gentle drift has already changed our planet: days have grown longer, and the Moon now sits much farther away than when it first formed. If this continues, future generations will see a slightly smaller Moon in the sky, and total solar eclipses may one day become a thing of the past.

Source:
https://www.bbc.com/future/article/20230303-how-the-moon-is-making-days-longer-on-earth

6

A Planet Made of Diamonds - Meet the Cosmic Gem

Imagine a world twice the size of Earth, orbiting so close to its star that a year lasts just 17 hours and surface temperatures soar to a scorching 4,400°F (2,400°C).

Welcome to 55 Cancri e, a super-Earth about 41 light-years away that's as extreme as it is fascinating. Early research suggested this fiery planet might be packed with diamonds making up a huge chunk of its interior due to its carbon-rich makeup and crushing pressures.

Thanks to the James Webb Space Telescope, we now know 55 Cancri e sports a thick, volcanic atmosphere, likely fed by lava flows on its blistering surface. While scientists are still unraveling exactly how much diamond lies beneath, this planet is a dazzling reminder that the universe is full of strange, extraordinary worlds, some of which might just sparkle like cosmic jewels.

Source:
https://en.wikipedia.org/wiki/55_Cancri_e

7

NOT ALL SPACE IS SILENT

If you were floating in deep space, you wouldn't hear a thing. There's no air or other material for sound waves to travel through, so yelling or clapping would be met with total silence. This is because sound requires a medium like air, water, or solid objects to carry its vibrations, and the vacuum of space simply doesn't provide that.

However, space isn't a perfect vacuum everywhere. In rare places where clouds of gas or plasma exist—like inside galaxy clusters—sound waves can ripple through the material. For example, NASA recently "heard" the eerie, low-pitched sound of waves moving through the hot gas around a supermassive black hole in the Perseus galaxy cluster, by converting those waves into audio humans can hear.

So while space is almost entirely silent for human ears, the universe does have a few places where sound, in a very different form, can echo across the cosmos.

Source:
https://theconversation.com/why-isnt-there-any-sound-in-space-an-astronomer-explains-why-in-space-no-one-can-hear-you-scream-217885

8

Jupiter's Red Spot: A Shrinking Giant Storm

Jupiter's Great Red Spot is the largest and most famous storm in our solar system, a swirling anticyclonic tempest that's been raging for at least 150 years, and possibly much longer.

NASA Hubble Space Telescope observations in 2024–2025 reveal that the Red Spot is now smaller than ever recorded, measuring about 10,250 miles (16,500 kilometers) across, just a bit larger than Earth, whereas in the late 1800s it was wide enough to fit three Earths side by side. The storm's shape has also changed, becoming more circular and oscillating in size and brightness, almost like a cosmic stress ball.

Scientists are still unraveling the reasons behind its ongoing shrinkage and dynamic behavior, but the Great Red Spot remains a powerful symbol of Jupiter's turbulent atmosphere and the mysteries of planetary weather.

Source:
https://science.nasa.gov/missions/hubble/hubble-shows-that-jupiters-great-red-spot-is-smaller-than-ever-seen-before/

9

Neutron Stars - Tiny Titans of the Cosmos

Neutron stars truly are the heavyweights of the cosmos—small in size, but unimaginably massive. They are so dense that a single teaspoon of their material would weigh about 6 to 7 billion tons on Earth, roughly the mass of a mountain.

These stars are only about 20 kilometers (12 miles) wide, yet they pack more mass than our Sun into that tiny space. Formed when massive stars explode in supernovae, neutron stars are so compact that their atoms are crushed together, creating matter denser than anything found elsewhere in the universe. Their gravity is so intense that they warp space and time, and their magnetic fields are trillions of times stronger than Earth's.

Source:
https://phys.org/tags/neutron+stars/

10

There Might Be More Planets Than Stars

Imagine looking up at the night sky and realizing that for every star twinkling above, there could be not just one—but several planets orbiting it.

Scientists now believe that the Milky Way galaxy is packed with at least as many planets as stars, and quite possibly many more. Some estimates suggest there could be trillions of planets out there, including countless Earth-sized worlds in the "habitable zone" where life as we know it might exist. Even stranger, many of these planets don't orbit any star at all—they drift alone through the vastness of space.

This staggering abundance of planets opens up exciting possibilities about the diversity of worlds waiting to be discovered and the potential for life beyond our solar system.

Source:
https://science.nasa.gov/exoplanets/

Explore more great reads from Third Rock Publishing

The Happiness Coach's Guide to a Fun Retirement,
by John Zakour
https://mybook.to/T4qXqA

A practical, uplifting guide to making your retirement years your happiest yet. Packed with humor, wisdom, and actionable tips. This book is your companion for creating a joyful, meaningful next chapter. John Zakour is a Certified Happiness and Wellness Coach, and was a gag writer for The Simpsons, Rug Rats, and continues many other humorous media adventures.

Time to Shine: Fun Short Stories - Retirement Gifts for Men and Women
by Third Rock Publishing
https://mybook.to/FJp2NO

A delightful collection of light-hearted, inspiring and uplifting short stories about men and women celebrating the adventures and surprises of retirement life. Perfect as a gift or for anyone looking to add a little sparkle to their day.

About Third Rock Publishing

Third Rock Publishing is a small, independent press dedicated to sharing authentic, uplifting, and insightful information and stories for curious and passionate readers.

We believe great books can spark joy, inspire growth, and foster genuine connection—whether they're grounded in science, personal development, or the adventures of everyday people. Every title we publish is meticulously researched and thoughtfully written.

We're committed to quality and integrity. If you'd like to connect, share feedback, or suggest a topic for a future book, we'd love to hear from you at info@thirdrockpublishing.com

Maggie

Editor

Third Rock Publishing
Email: info@thirdrockpublishing.com

www.ingramcontent.com/pod-product-compliance
Lightning Source LLC
LaVergne TN
LVHW022038200926
848481LV00003B/717